CLASSICS OF SCIENCE

Volume VI

Radiochemistry
and
the Discovery of Isotopes

CLASSICS OF SCIENCE SERIES

under the General Editorship of

Gerald Holton

Professor of Physics, Harvard University

ALREADY PUBLISHED

CLASSICS IN THE THEORY OF CHEMICAL COMBINATION
edited by O. Theodor Benfey

THE DISCOVERY OF RADIOACTIVITY AND TRANSMUTATION
edited by Alfred Romer

THE DEVELOPMENT OF HIGH-ENERGY ACCELERATORS
edited by M. Stanley Livingston

CLASSICS IN COORDINATION CHEMISTRY, PART I
edited by George B. Kauffman

SOURCES OF QUANTUM MECHANICS
edited by B. L. van der Waerden

RADIOCHEMISTRY AND THE DISCOVERY OF ISOTOPES
edited by Alfred Romer

Classics of Science, Volume VI

Radiochemistry and the Discovery of Isotopes

Edited, with commentary and an introductory historical essay, by

ALFRED ROMER

Henry Priest Professor of Physics
St. Lawrence University

DOVER PUBLICATIONS INC.

NEW YORK

Published in Canada by General Publishing Company, Ltd., 30 Lesmill Road, Don Mills, Toronto, Ontario.

Published in the United Kingdom by Constable and Company, Ltd., 10 Orange Street, London WC 2.

Radiochemistry and the Discovery of Isotopes (Volume VI in the Dover "Classics of Science" series) contains twenty-six papers published for the first time in collected form by Dover Publications, Inc., in 1970. Eighteen of these papers were translated from French or German by Professor Alfred Romer; the remaining eight, in English, have been reprinted without change from the original journals. Professor Romer has also provided a Preface, an Historical Essay, and editorial commentary on the papers.

Standard Book Number: 486-62507-9
Library of Congress Catalog Card Number: 74-91273

Manufactured in the United States of America

Dover Publications, Inc.
180 Varick Street
New York, N.Y. 10014

General Editor's Foreword

This is the sixth volume in a new publishing program entitled *Classics of Science*. Each volume is a collection of fundamental essays and other basic original articles in a certain field of science, presented in the sequence of its development, together with an introduction, commentary, and clarifying notes by the scholar responsible for the selection of the papers. All areas of science are to be included as the series grows. The titles of volumes range from *The Discovery of Radioactivity and Transmutation* to *The Theory of Evolution* (*1830 to Modern Genetic Theory*). Thus, in time, we shall have here a convenient network of roads to take us to the often inaccessible sources of the great rivers of science, through its widely visible mountain ranges, its renowned battlefields, and perhaps its little-known but nevertheless choice vineyards.

The articles follow the original texts verbatim, and (with rare and clearly marked exceptions) are printed in full rather than in excerpt or in edited and abbreviated versions; they may therefore be used in lieu of the original publications. Foreign-language articles are carefully translated. In short, both the interested layman and the historian or scientist should feel assured that he is reading these documents as they were intended to be read, and in the version in which they made their original contribution.

Some of these aims are similar to those of a distinguished German series of republications, begun in 1889 under the direction of the chemist and philosopher, Wilhelm Ostwald, and reaching some 400 titles before it was discontinued. There are, however, important distinctions between these two series. Ostwald's series was composed of republications of single essays, whereas here each book puts together several essays, each in the context of the others, in order to trace the development of a whole field of study. Hence there will be more connective commentary in this series, and, most importantly, the commentary can now benefit from advances made in the last three-quarters of a century, both in science and in historical scholarship.

Still, many of the aims cited in the original announcement of Ostwald's *Klassiker der exakten Wissenschaften* are again applicable:

The great progress which the sciences have made in our time, as is generally acknowledged, is in good measure owing to the cultivation and wide application of teaching methods such as demonstration lectures and laboratories. These arrangements are indeed very successful in imparting knowledge of the present state of science. But some of the most eminent persons with widest vision have frequently felt compelled to point out a deficiency that mars all too often our contemporary scientific education. This is the absence of a sense of history, and the lack of knowledge concerning those great contributions on which the edifice of science rests.

Although few instructors in scientific subjects would fail to make reference to those foundations when the occasion arises, such references remain, nevertheless, generally ineffective because the source materials of science are rarely accessible. They can be obtained only in the larger libraries, and then only in single copy, so that the student is all too easily discouraged from pursuing the lead.

This lack is to be remedied through the publication of the *Klassiker der exakten Wissenschaften*. In convenient format and at reasonable prices, the seminal publications of all the exact sciences will be made available to instructors and students alike. The publisher hopes to create thereby both a teaching aid that gives life and depth to the study of science, as well as a significant research tool; for in those fundamental writings lie not only the seeds which have, in the meantime, developed and borne fruit, but also unnumbered other leads that await development. To those whose study and research lie in the sciences, these works offer an inexhaustible source for stimulation and for the advancement of ideas.

Today, no less than when these words were written, the proper orientation of science is toward the future. Yet, the uses of the scientific past are also becoming clearer—not the least being the continued application of Maxwell's memorable dictum on the didactic value of the study of historical accounts and of original works in science, found in the preface to the *Treatise on Electricity and Magnetism*: "It is of great advantage to the student of any subject to read the original memoirs on that subject, for science is always most completely assimilated when it is in the nascent state. . . ."

This series of connected essays on single topics of science will help us to remember that in the development of each field the overriding characteristic of scientific growth is its continuity. If the quality of science today is so good, it is to a large extent because science in the past was so good.

Gerald Holton
Jefferson Laboratory
Harvard University

Preface

This book, and its previously published companion volume in the *Classics of Science* Series, *The Discovery of Radioactivity and Transmutation*, have been planned to illustrate the main lines of development of the science of radioactivity between 1896 and 1913. I have made an effort in each volume to present a continuous narrative through the papers reprinted and the editorial commentary which links them. The choice of papers to be included was made with that end in view. The initial discoveries are here, and so are the final statements of achievement; but so also are the slow advances between them, some moving in a direction as yet unspecified, others oriented to test a promising but speculative hypothesis.

For some years I have felt that histories of the sciences should be organized around the great clarifying principles which unify and render intelligible the scattered phenomena of experiment and observation. When I began my study of the history of radioactivity, I saw such a principle in the transformation theory, and as I continued, I found a second in the theory of isotopes, which had been proposed (as I discovered) simply on radiochemical evidence. I have made these the subjects of the two volumes which I have undertaken to edit for this Series.

By the time of publication of the first volume, it was evident that the narrative form of reprinted paper and editorial commentary was somewhat cumbersome. It has seemed useful to add to the second volume a historical essay which would set out in relatively brief and connected form the history which the two collections of reprinted papers exemplify. It may be worth adding that as history this essay is incomplete (and indeed all written history must be), since I have omitted topics which did not bear directly on the development of its organizing principles. There is little mention here of the later work of the Curies, only the slightest mention of Bragg's analysis of alpha-particle ranges, and nearly total neglect of the discovery in Rutherford's laboratory of the atomic nucleus. I have left them out not because they were unimportant, but because they contributed little to the themes of this particular story. It has been the same with the discovery of the branching products and the detection of Po^{214} and Po^{216} in Rutherford's laboratory.

The papers I have chosen to reprint are those which seemed to me important and interesting. I have tried to show the initial, puzzling discoveries which had to be understood, the remarkably slender bases in personal experience out of which hypotheses were formulated, and the success with which those hypotheses were subsequently applied. I have included also a few polemical papers from the great polonium controversy, which may perhaps be considered "bad" science, but which illustrate rather clearly the scientific value of an organizing principle, in this case the transformation theory.

The brief editorial commentary between the reprinted papers is in each case enclosed in square brackets for identification; that it occasionally overlaps with the general historical essay is unavoidable and is not thought to be a handicap for the scholar or student who will use these volumes. With a single exception (which is noted) in each volume, the papers presented here are reprinted in full—those in English in the original text, those which appeared in French or German in my own translation.

Since the preparation of the first draft of this volume, our knowledge of the history of radioactivity had been considerably enlarged by the researches of Dr. Lawrence Badash. Although many of the results of his work are yet to be incorporated into current publications, such as this present volume, I believe that they will illuminate a number of the problems discussed here.

I am grateful to the many people who have assisted me in this project. Let me single out first the whole profession of librarians, who have so organized the world over which they preside that its riches open themselves to inquiry. Let me mention also my wife, Emily, who provided much of the original impetus for these studies but did not live to see them finished. Of those who have been concerned with this book in particular, I must speak of three by name. I am grateful to Gerald Holton, the General Editor of this Series, for giving me a free hand in the organization and presentation of my material, and for his patient insistence on intelligibility. I owe also a large debt of gratitude to Mrs. Joyce Clintsman and Mrs. Joan Mousaw, who between them have typed the manuscript with unfailing patience and accuracy.

A. R.

Canton, New York
September 1969

Contents

List of Illustrations

Historical Essay

by

ALFRED ROMER

The Science of Radioactivity, 1896–1913
Rays, Particles, Transmutations, Nuclei and Isotopes

by Alfred Romer

Introduction. From time to time in the sciences a new field may open which shortly before had been inconceivable. An unnoticed phenomenon comes to attention, a novel concept is formulated, and what had previously been matter for speculation is brought within the range of experimental research. Just such a development occurred in the field of radioactivity during the first years of the twentieth century, a development which brought physics from a grudging half-belief in the existence of atoms to a clear conception of their internal structure and to the first steps in nuclear physics.

The key operation—as it is still for nuclear physics—was the detection and measurement of ionizing radiation. The history of radioactivity begins with the history of x rays. The discovery of one penetrating, ionizing radiation led to another which occurred as a spontaneous emission from uranium. The recognition of these rays as characteristic of the element uranium led first to the discovery of hitherto unknown, ray-emitting elements, and later to the astonishing conclusion that all such ray-emitting elements were spontaneously being transmuted, precisely in the alchemical sense. From this knowledge, two new problems arose: to trace the genetic lines of the radioactive substances, and then, by determining their chemical nature, to reconcile their multiplicity with the limited room in the periodic table. It was in the struggle with this last difficulty that the existence of isotopes was first recognized.

The problems of establishing the transmutation theory and of working out the chemical identities of the radioactive elements provide the themes for two volumes in the *Classics of Science* series, *The Discovery of Radioactivity and Transmutation*, and the present *Radiochemistry and the Discovery of Isotopes* (to be referred to in this historical essay and in my footnotes and commentary throughout the volume as Vol. I and Vol. II respectively). Those problems do not of course constitute the whole history of radioactivity, and it will be useful to touch here on such other matters as the nature of the rays emitted and their use as probes in exploring the structure of the atom.

Uranium, 1896–1897. Radioactivity began as a branch of optics. When Röntgen startled the world in the last days of 1895 with his x-ray pictures of hidden things, he had used tubes in which the target for the cathode-ray beam was simply the glass envelope. He had reported then quite accurately that the source of the x rays lay in the fluorescent spot where the cathode rays struck. This led easily to the speculation that the same mechanism produced both x rays and visible light, and that x rays might form a hitherto unnoticed part of all fluorescence, by whatever means it was excited.[1]

Henri Becquerel, Professor at the Muséum d'Histoire Naturelle in Paris and at the École Polytechnique, was experienced in dealing with light, and he welcomed the possibility of producing x rays by purely optical means. By the end of February, 1896, he succeeded in obtaining a photographic exposure through a protective wrapping of black paper from a crystal of fluorescent potassium uranyl sulfate while it lay for several hours exposed to sunshine[2] (Paper 1 in Vol. I).

Actually his working hypothesis was false, and most of the circumstances which appeared to corroborate it irrelevant, but Becquerel was an intuitive experimenter who could make the most of the unexpected. While still holding firmly to his original ideas, by May he had managed to eliminate the irrelevancies and to link the penetrating radiation to the simple presence of uranium.

Within a week of his first success, he knew that potassium uranyl sulfate could emit the penetrating rays in darkness although that particular substance required constant illumination to maintain its fluorescence of visible light[3] (Paper 2 in Vol. I). Within another week, he discovered that like x rays, these penetrating rays possessed the power to discharge electrified bodies. He found that in addition to potassium uranyl sulfate, other fluorescent compounds of uranium emitted the rays, as was rather to be expected, but in darkness as well as in the light, as was not. Unexpectedly again, fluorescent compounds of calcium and zinc failed to produce them.[4]

---

[1] H. Becquerel, "Recherches sur une propriété nouvelle de la matière," *Mémoires de l'Académie des Sciences, Paris,* 1903, *46,* 360 pp., 13 plates.

H. Poincaré, "Les rayons cathodiques et les rayons Röntgen," *Revue générale des sciences pures et appliquées,* 1896, *7*: 52–59.

[2] H. Becquerel, "Sur les radiations émises par phosphorescence," *C. R. Acad. Sci., Paris,* 1896, *122*: 420–421.

[3] H. Becquerel, "Sur les radiations invisibles émises par les corps phosphorescents," *ibid.*, pp. 501–503.

[4] H. Becquerel, "Sur quelques propriétés nouvelles des radiations invisibles émises par divers corps phosphorescents," *ibid.*, pp. 559–564.

ANTOINE HENRI BECQUEREL (1852-1908)
At the Age of Thirty-Six
[From *Die berühmten Erfinder, Physiker und Ingenieure.* Cologne: Aulis Verlag Deubner & Co.]

HENRI MOISSAN (1852-1907)

Becquerel would have liked to discover what portion of the optical spectrum excited his rays, but this was impossible until he could obtain a quiescent specimen which he might excite. He set aside some of his preparations in darkness so that they might exhaust themselves, but day after day they continued to pour out their rays. Non-fluorescent compounds of uranium, which had no power to store and re-emit the energy of visible light, nevertheless gave out the penetrating rays. Fluorescent uranyl nitrate, isolated in darkness, dissolved in its own water of crystallization, and re-crystalized there, still managed to produce the rays[5] (Paper 3 in Vol. I). In fact a solution of uranyl nitrate, which was not itself fluorescent, emitted them as freely as did the solid crystals.[6]

The one constant factor in every successful experiment had been the presence of uranium, whether in fluorescent or non-fluorescent compounds, and that suggested that uncombined, metallic uranium might also produce the rays. A fluorescent metal would be an oddity; if Becquerel were to discover one, he wanted to be perfectly sure of his ground. The usual reduction process for uranium produced not a pure metal but a mixture of metal and carbide; however he knew that Henri Moissan of the École Supérieure de Pharmacie had immediate hopes of a complete reduction with his new electric-arc furnace. By May, Moissan had succeeded,[7] and with a disc of his metallic uranium, Becquerel was able to obtain photographic exposures and the discharge of a gold-leaf electroscope with something like four times the intensity he had first observed with potassium uranyl sulfate[8] (Paper 4 in Vol. I).

Becquerel continued to work with his "uranic rays" into the spring of 1897, but to very little purpose. In particular, the source of energy which supported them remained a mystery, and a persistently troubling mystery, since the preparations which had lain isolated in darkness now for a year still gave out their penetrating radiation.[9] In the fall of 1897 he turned his attention to magneto-optics and the newly-discovered Zeeman effect.[10]

---

[5] H. Becquerel, "Sur les radiations invisibles émises par les sels d'uranium," *ibid.*, pp. 689–694.

[6] H. Becquerel, "Sur les propriétés différents des radiations invisibles émises par les sels d'uranium, et du rayonnement de la paroi anticathodique d'un tube de Crookes," *ibid.*, pp. 762–767.

[7] H. Moissan, "Préparation et propriétés de l'uranium," *ibid.*, pp. 1088–1093.

[8] H. Becquerel, "Émission de radiations nouvelles par l'uranium métallique," *ibid.*, pp. 1086–1088.

[9] H. Becquerel, "Sur diverses propriétés des rayons uraniques," *ibid.*, 1896, *123*: 855–858; "Recherches sur les rayons uraniques," *ibid.*, 1897, *124*: 438–444; "Sur la loi de décharge dans l'air de l'uranium électrisé," *ibid.*, pp. 800–803.

[10] H. Becquerel, "Sur une interprétation applicable au phénomène de

Polonium and Radium, 1897–1898. Two of Becquerel's successors (who had been attracted to his experiments simply by reading about them), began with an interest in the rays. Gerhard Carl Schmidt of Erlangen supposed that their power to discharge electrified bodies might be a manifestation of the photoelectric effect. He found it was not.[11] Marie Curie in Paris, wife of a professor at the École Municipale de Physique et de Chimie Industrielles, attempted to change their intensity by heating a disc of Moissan's metallic uranium, by irradiating it with light, and by exposing it to x rays. All of these efforts failed.[12] Then to both of them, independently and simultaneously, came the realization that the emission of penetrating rays might be a specific, elementary property of the atoms of uranium, and that the same property might be possessed by the atoms of other elements. Each found such an element in thorium.[13]

Marie Curie detected the characteristic rays by their electrical effects, using the Curie electrometer which her husband Pierre and his brother Jacques had designed. That gave her the power to compare intensities, and she was astonished to notice that pitchblende, a natural oxide of uranium, had a higher activity of radiation than the pure, uncombined metal. The discovery stimulated her to test other uranium preparations, the refined compounds of the chemist as well as minerals. For the chemical specimens, she found the intensity of the rays to be consistently proportional to their uranium content. The

---

Faraday et au phénomène de Zeeman," *ibid.*, 1897, *125*: 679–685; "Remarques sur la polarisation rotatoire magnétique et la dispersion anomale, à l'occasion d'une expérience nouvelle de MM. D. Macaluso et O.-M. Corbino," *ibid.*, 1898, *127*: 647–651; "Sur la dispersion anomale et le pouvoir rotatoire magnétique de certaines vapeurs incandescentes," *ibid.*, pp. 899–904; "Sur la dispersion anomale de la vapeur de sodium incandescente, et sur quelques conséquences de ce phénomène," *ibid.*, 1899, *128*: 145–151.

H. Becquerel and H. Deslandres, "Contribution à l'étude du phénomène de Zeeman," *ibid.*, 1898, *126*: 997–1001; "Observations nouvelles sur le phénomène de Zeeman," *ibid.*, *127*: 18–24.

[11] E. Wiedemann and G. C. Schmidt, "Ueber Luminescenz von festen Körpern und Festen Lösungen," *Ann. Phys. Lpz* [3], 1895, *56*: 201–254, esp. pp. 241–250.

[12] "Étude des 'Carnets de laboratoire,'" ed. I. Joliot-Curie, pp. 106–107, in Marie Curie, *Pierre Curie*, Paris: Denoël, 1955.

[13] G. C. Schmidt, "Ueber die vom Thorium and den Thorverbindungen ausgehende Strahlung, "*Verhandlungen der physikalische Gesellschaft zu Berlin*, 1898, *17*: 13, 14–16; "Ueber die von den Thorverbindungen und einigen anderen Substanzen ausgehende Strahlung," *Ann. Phys., Lpz* [3], 1898, *65*: 141–151; "Sur les radiations émises par le thorium et ses composés," *C. R. Acad. Sci., Paris*, 1898, *126*: 1264.

Mme. Sklodowska Curie, "Rayons émis par les composés de l'uranium et du thorium," *C. R. Acad. Sci., Paris*, 1898, *126*: 1101–1103.

"Carnets de laboratoire," *op. cit.*, pp. 108–112.

PIERRE CURIE (1859-1906)

MARIE SKLODOWSKA CURIE (1867-1934)

EUGÈNE DEMARÇAY (1852-1904)
[From *Torchbearers of Chemistry* by Henry M. Smith. New York: Academic Press, 1949.]

radiation from her minerals was regularly more intense. Thus it was plausible, as she pointed out, that these minerals might contain another element even more active in radiation than uranium (Paper 1 in Vol. II).

By this time, in April 1898, Pierre Curie had been drawn into the investigation. Together they began a chemical analysis of pitchblende, drawing on the technical knowledge of Gustave Bémont, *chef des travaux* for chemistry at the school. Their procedure was that of the undergraduate analysis (as it was taught then), but it needed no particular sophistication to recognize what they hoped to find, a substance which was neither uranium nor thorium and yet was "radioactive." (This was a word they had coined to describe the steady emission of penetrating rays.) By July they had found one. It accompanied bismuth in their analytical scheme, but it could be distinguished from bismuth by its radioactivity and by the greater volatility of its sulfide. Slight though this information was, it showed that their material corresponded to no known substance, and the two physicists ventured to claim the discovery of a new element which they named *polonium* for the then nonexistent nation of Marie Curie's birth[14] (Paper 2 in Vol. II).

During the remainder of 1898, the Curies and Bémont struggled with another radioactive substance which accompanied barium, and here their problem was not simply to recognize the material but to concentrate it from the barium salts with which it was mixed. The procedure which proved effective was a long series of partial precipitations, from which they obtained samples that were at first 60 times, then 227, then 600 times as active as an equal quantity of uranium. If this substance also was a new element, given the intensity of its radiation they could imagine no name more appropriate than *radium*.

From time to time, as the concentration progressed, the Curies had brought their richest specimens to the spectroscopic laboratory of Eugène Demarçay. He finally obtained, along with the spectra of barium and of predictable impurities, one line in the near ultraviolet that he could assign to no known element and that grew steadily stronger as the radioactivity of the samples increased. This was evidence of the right kind; at the end of December the Curies announced the discovery of their second radioactive element[15] (Papers 3 and 4 in Vol. II).

---

[14] P. Curie and Mme. S. Curie, "Sur une substance nouvelle radio-active, contenue dans la pechblende," *C. R. Acad. Sci., Paris*, 1898, *127*: 175–178.
"Carnets de laboratoire," *op. cit.*, pp. 112–115.

[15] P. Curie, Mme. P. Curie and G. Bémont, "Sur une nouvelle substance fortement radio-active, contenue dans la pechblende," *C. R. Acad. Sci., Paris*, 1898, *127*: 1215–1218.

To complete the discovery, they needed to determine the atomic weight of their new element, and the purchase of pitchblende to accumulate enough radium promised to be prohibitively expensive. Fortunately, Professor Eduard Suess of the University of Vienna was a *Correspondent* of the Académie des Sciences. Through his intercession, they were able to obtain from the uranium refinery operated by the Austrian government at Joachimsthal in Bohemia 100 kilograms of the residues, which presumably still contained radium, if they would pay only the shipping charges.[16] (Commercially, uranium was used as a coloring agent in glass and glazes and to some extent as an alloy constituent in steel.)

Rays and Ions, 1896–1900. During the next two years, it was discovered that at least some of the rays from radium were corpuscular, consisting of streams of subatomic particles which carried negative charges and moved at speeds approaching that of light. This was sufficiently remarkable, even for radium, since the possible existence of subatomic particles had been demonstrated so recently that they were hardly yet credible to the scientific community.

This line of research went back to the first frantic weeks after the discovery of x rays when everyone who possessed a cathode-ray tube was making skeletal photographs. It had been noted then that x rays transformed air temporarily from a non-conductor to a conductor of electricity. This phenomenon was vigorously investigated during the spring and summer of 1896 by J. J. Thomson, Cavendish Professor of Experimental Physics at Cambridge University, and Ernest Rutherford, a research student of his from New Zealand. Thomson thought in atomic terms (although perhaps he would have said "molecular") about gases and about electricty. The theory of conducting gases which he developed, and which their studies supported, was an ionic one.

He supposed that the x rays operated to create pairs of oppositely charged ions from neutral molecules of a gas. What appeared as the conduction of charges across the space from one electrode to the other was rather the simultaneous discharge of both electrodes by ions drawn toward them from the body of the gas. With sufficiently strong fields between the electrodes, these ion currents became independent of the field, an indication that the ions were being withdrawn as rapidly as the x rays produced them. Such a "saturation current" could serve then as a reliable index of the intensity of the ionizing rays.[17]

E. Demarçay, "Sur le spectre d'une substance radio-active," *ibid.*, p. 1218. "Carnets de laboratoire," *op. cit.*, pp. 115–119.

[16] M. Curie, *Pierre Curie*, Paris: Denoël, 1955, p. 62.

[17] J. J. Thomson and E. Rutherford, "On the Passage of Electricity through Gases exposed to Röntgen rays," *Phil. Mag.* [5], 1896, *42*: 392–407.

ERNEST RUTHERFORD (1871-1937)

It is interesting to notice that the Curies were unaware of this work, and had to validate the method empirically when they began using ionization currents to measure their ray-intensities.

In the fall of 1896, Thomson withdrew from the ionization studies, but Rutherford continued with them for the next two years. He measured the rate at which ions would recombine when they were not removed from the gas. He invented means for separating positive ions from negative without discharging them on electrodes. He assessed the speed with which they moved in an electric field. He studied the degree of ionization which the rays produced in different gases, and as a related phenomenon, the extent to which those gases absorbed the rays. As he mastered the details of gaseous ionization, Rutherford moved from x rays as the agent to ultraviolet light and then to uranium. The paper on ionization by uranium, which he published in 1899, might almost be regarded as the masterpiece by which this research student qualified for admission to the guild of working physicists. It is overelaborate, as were its medieval counterparts, yet for its relevance, thoroughness, and logic it forms an admirable example of the craft of experimental physics.[18]

One small part of this thoroughgoing paper is of particular interest here. While he was studying the absorption of uranium rays by metal foils, Rutherford discovered that they were not homogeneous, but consisted of two distinct sets, which in a purely formal way he named alpha and beta rays. The alpha rays were not penetrating, they could be stopped for example by a sheet of heavy paper, but they produced 95 per cent of the ionization in his apparatus. The beta rays, which produced the remaining 5 per cent, were about as penetrating as the x rays he was accustomed to.

In the meanwhile, Thomson had shifted his attention to cathode rays. They formed a well-known and puzzling part of the high-voltage discharge through gases at low pressures. They appeared as blue streamers which spread out from the negative electrode, and which produced x rays, as Röntgen had shown, when they played on any solid obstacle.

Perrin had recently shown that they carried negative charge into a Faraday cage,[19] and this bore out Thomson's firm opinion that the

---

[18] E. Rutherford, "On the Electrification of Gases exposed to Röntgen Rays, and the Absorption of Röntgen Radiation by Gases and Vapours," *Phil. Mag.* [5], 1897, *43*: 241–255; "The Velocity and Rate of Recombination of the Ions of Gases exposed to Röntgen Radiation," *ibid.*, *44*: 422–440; "The Discharge of Electrification by Ultra-Violet Light," *Proceedings of the Cambridge Philosophical Society*, 1898; *9*, 401–416; "Uranium Radiation and the Electrical Conduction produced by it," *Phil. Mag.* [5], 1899, *47*: 109–163.

[19] J. Perrin, "Nouvelles propriétés des rayons cathodiques," *C. R. Acad. Sci., Paris*, 1895, *121*: 1130–1134.

bluish lines marked the path of swift-moving ions, which he thought were probably derived from the gas in the tube. Devising experiments to test his ideas, he soon demonstrated that in a uniform magnetic field the cathode-ray streamers curved in a circular arc, and thus had the same trajectory as moving ions. By measuring simultaneously the rate at which a cathode-ray beam transported charge and energy, and combining this information with the curvature of the rays in his magnetic field, Thomson calculated that the ions whose existence he postulated must carry some 10^7 electromagnetic units of charge per gram of mass. For hydrogen ions in electrolysis, the corresponding charge-to-mass ratio was 10^4. The cathode-ray ions were evidently quite different from hydrogen ions. If one could believe that their charges were comparable, then they must be remarkably small, must indeed be sub-atomic particles.

By the summer of 1897, he had carried out deflection experiments in a transverse electric field to confirm the high value of the charge-to-mass ratio, but he needed two years more to prove conclusively that this indicated a sub-atomic mass. Progressing step by step, by the end of 1898 he had succeeded in measuring the absolute value of the average charge carried by each separate ion in a gas exposed to x rays. By the fall of 1899, he had tied the two lines of investigation together by means of measurements on the photoelectric ions (of unknown nature) produced when ultraviolet light shone on a metal plate. The charge-to-mass ratio of the photoelectric ions was that of the cathode-ray ions, the charge they carried that of the ions liberated by x rays. In that same year J. S. Townsend, another of his research students, completed the proof by demonstrating that the x-ray ions carried the same charge as did the hydrogen ions in electrolysis. The two charge-to-mass ratios might then be legitimately compared, and thus the cathode-ray ions, the photoelectric ions, and in addition the ions evaporated from a white-hot filament of carbon, all represented a new variety of negatively charged subatomic particle. The masses of these particles would be something like a thousandth of the mass of a hydrogen ion, and to Thomson it seemed plausible that they might form the undifferentiated, basic stuff of matter. For this reason, he named them *corpuscles*,[20] a name which was soon replaced in popular favor by *electron*.

By 1899, when the work of Thomson and others on cathode rays had begun to attract attention, radium was becoming generally available. Becquerel had received some from the Curies, with whom

[20] J. J. Thomson, "Cathode Rays," *Electrician*, 1897, *39*: 104–109; "Cathode Rays," *Phil. Mag.* [5], 1897, *44*: 293–316; "On the Charge of Electricity carried by the Ions produced by Röntgen Rays," *Phil Mag.* [5], 1898, *46*: 528–545; "On the Masses of the Ions in Gases at Low Pressures," *ibid.*, 1899, *48*: 547–567.

he had struck up an acquaintance. In Germany Friedrich Giesel, an active chemist and the director of Buchler und Compagnie of Braunschweig, a firm of quinine manufacturers, had managed to concentrate both radium and polonium and was generous in giving away specimens (Paper 5 in Vol. II). In the neighboring city of Hannover, the firm of de Haën put radium preparations on the market.[21]

In Wolfenbüttel, near Braunschweig, J. Elster and H. Geitel found that the rays from radium discharged electrified bodies more slowly when a magnetic field was applied. They did not think in ionic terms, and thus attributed the phenomenon to a reduction in conductivity of the air. As Giesel immediately showed, it arose actually from a transverse deflection of the rays. In Vienna, Stefan Meyer and Egon Ritter von Schweidler (with a little prompting by correspondence from Giesel) managed to confirm the occurrence of such a deflection and pointed out that the rays curved in the same direction as would a current of negative electricity flowing outward from the radium.[22]

Becquerel resumed his studies in radioactivity by reviewing some of the recent experiments with uranium, then turned his attention to the rays from radium.[23] By the late fall, he was trying the effects of a magnetic field, and since he thought habitually in optical terms, he began with the field parallel to the line of the rays. After the announcements by Meyer and von Schweidler, he shifted to a transverse field, which was certainly better adapted to the study of ions.

By the end of the year, Pierre Curie had also begun to investigate the rays. Combining in his research both magnetic-deflection and absorption studies, he distinguished two types of radiation in the emission from radium. The rays most easily absorbed, those that disappeared after 6.5 cm of travel through air, he found to be entirely unaffected by the magnetic field. The more penetrating rays, those

---

[21] F. Giesel, "Einiges über das Verhalten des radioactiven Baryts und über Polonium," *Ann. Phys., Lpz* [3], 1899, *69*: 91–94; "Über Radium und Polonium," *Phys. Z.*, 1899–1900, *1*: 16–17.

E. de Haën, "Ueber eine radioaktive Substanz," *Ann. Phys., Lpz* [3], 1899, *68*: 902.

[22] J. Elster and H. Geitel, "Weitere Versuche an Becquerelstrahlen," *Ann. Phys., Lpz* [3], 1899, *69*: 83–90.

F. Giesel, "Ueber die Ablenkbarkeit der Becquerelstrahlen im magnetischen Felde," *ibid.*, pp. 834–836.

S. Meyer and E. R. von Schweidler, "Über das Verhalten von Radium und Polonium im magnetischen Felde," *Phys. Z.*, 1899–1900, *1*: 90–91, 113–114.

[23] H. Becquerel, "Note sur quelques propriétés du rayonnement de l'uranium et des corps radio-actifs," *C. R. Acad. Sci., Paris*, 1899, *128*: 771–777; "Recherches sur les phénomènes de phosphorescence produits par le rayonnement du radium," *ibid.*, *129*: 912–917.

which remained when the radium was placed at a distance from the magnet, proved to be easily deflected when they reached the magnetic field.

After these preliminaries, in the early months of 1900, the same sequence of experiments which had revealed the nature of the electron was repeated with the magnetically-deviable rays from radium. In January, Becquerel demonstrated that their trajectory in a uniform magnetic field was the arc of a circle. By the beginning of March, the Curies had found that these rays carried negative charge into an insulated collector, and carried it away from an insulated specimen of radium. A week later Ernst Dorn of Halle, in Saxony, announced that he could deflect the same rays with an electric field. By the end of the month, Becquerel had made the experiments quantitative. From measured deflections in electric and magnetic fields, he showed that these deviable rays from radium were very probably negative ions, travelling at a speed of 1.6×10^{10} cm/sec and having a charge-to-mass ratio of 10^7. It appeared that Thomson's corpuscles played a part in radioactivity, as well as in gaseous discharges, in photoelectricity, and in thermionics.[24]

The Fugitive Radioactivities of Thorium, 1898–1900. In 1898, at the age of twenty-seven, Ernest Rutherford, Thomson's research student at Cambridge, was appointed Professor of Physics at McGill University in Montreal. It was a transatlantic post to be sure, but his predecessor, H. L. Callendar, had made a distinguished reputation there, and John Coxe, the chairman of the physics department, had given Rutherford to understand that research would be his primary responsibility. His study of uranium was now finished, but there would be value in a parallel treatment of the radiation from thorium, and this project caught the fancy of one of his new colleagues, R. B. Owens, the professor of Electrical Engineering.

Since Rutherford had used the oxide of uranium, Owens chose to work with thorium oxide, and immediately encountered difficulties. Far from remaining constant, its radiation appeared to change in

---

[24] H. Becquerel, "Influence d'un champ magnétique sur le rayonnement des corps radio-actifs," *C. R. Acad. Sci., Paris*, 1899, *129*: 996–1001; "Sur le rayonnement des corps radio-actifs," *ibid.*, pp. 1205–1207; "Contribution a l'étude du rayonnement du radium," *ibid.*, 1900, *130*: 206–211; "Déviation du rayonnement du radium dans un champ électrique," *ibid.*, pp. 809–815.

P. Curie, "Action du champ magnétique sur les rayons de Becquerel. Rayons déviés et rayons non-déviés," *ibid.*, pp. 73–76.

P. Curie and Mme. M. P. Curie, "Sur la charge électrique des rayons déviables du radium," *ibid.*, pp. 647–650.

E. Dorn, "Elektrostatische Ablenkung der Radiumstrahlen," *Abhandlungen der naturforschenden Gesellschaft zu Halle*, 1901, *22*: 6 pp. (separate pagination). Read at the session of Jan. 20, 1900.

intensity in response to events as irrelevant as the opening of a door. He traced the disturbance to air currents and obtained the necessary stability by enclosing the ion-collecting electrodes with the thorium oxide in a metal box. With these precautions, his research went ahead on the lines laid down by Rutherford with uranium.[25]

This, however, was as far as Owens cared to go. In the summer of 1899, Rutherford took over for his own research the puzzling sensitivity of thorium oxide. He decided to examine not the thorium oxide itself but the air which had blown past it, and there he found a continuing ionization which persisted for perhaps ten minutes. He convinced himself that this represented no alteration of the air, but rather a persistent re-ionization by radiation from some substance picked up as the air blew past the thorium oxide preparation, and picked up there even when that preparation had been carefully wrapped in paper.

This substance, which he referred to as an "emanation," was extraordinarily tenuous. It worked its way through the paper wrapping. It passed through every filter he set to intercept it. It appeared to be a radioactive gas, but with a radioactivity which was only temporary. The logic of Becquerel's experiments pointed to radioactivity as a permanent property of the atoms of uranium, and this assumption had underlain the discovery by Schmidt and the Curies of the radioactivities of thorium, polonium, and radium. Nevertheless, the radiation from Rutherford's emanation died down in a geometrical progression with time, halving its intensity every 60 seconds. Expressed analytically, that intensity could be written as

$$I = I_0 e^{-\lambda t} \tag{1}$$

where λ had the value of 1/86 when the time was expressed in seconds.

This equation implied that for a particular sample of emanation, the rate of decrease of intensity of its radiation was always proportional to the intensity itself at that particular instant; that is, that

$$\frac{dI}{dt} = -\lambda I. \tag{2}$$

This conclusion suggested an experiment to Rutherford. Suppose he were to put into his ionization chamber a sample of thorium oxide wrapped tightly in paper through which the emanation could diffuse at a steady rate. Then the intensity of radiation in the chamber, as measured by the ionization there, should increase steadily as the

---

[25] R. B. Owens, "Thorium radiation," *Phil. Mag.* [5], 1899, *48*: 360–387.

emanation accumulated, but also continue to decrease in proportion to its own intensity. That is to say that

$$\frac{dI}{dt} = k - \lambda I, \tag{3}$$

where k would depend on the rate of diffusion of the emanation through the paper. Consequently the degree of ionization in the chamber would be given by

$$I = I_f(1 - e^{-\lambda t}), \tag{4}$$

where I_f represented a maximum value toward which the ionization tended and λ was the same as in Eq. (1). When he tried the experiment, Rutherford found precisely what his analysis had predicted (Paper 5 in Vol. I).

While he was engaged with these and other details, he was plagued by an apparent failure of insulation in his ionization chamber. In actual fact, the trouble was less simple. The metal parts of that chamber had somehow become radioactive, and the supposed leakage was an extra ion current. Rutherford promptly named the phenomenon an "excited radioactivity," and then energetically proceeded to prove the name misleading. The new radioactivity was not in any sense "excited" by the rays from thorium oxide or by the proximity of that material. It appeared more closely related to the emanation, arising from an impalpable but material deposit on any surface which the emanation touched. The radiation remained the same whatever substrate he used, and although the deposit could neither be blown away nor driven off by heating (its continuing presence being shown by its radioactivity), it would dissolve in suitable acids and remain on a glass dish when the solution was evaporated.

The particles of emanation behaved like ordinary matter and were not affected by the presence of an electric field. In contrast, the carriers of the excited radioactivity appeared to be positively charged since they could be concentrated on any negatively charged body. Like the particles of emanation, however, they lost radioactivity in a geometric progression with time, although at a considerably slower rate. They required 11 hours for each halving of intensity, which gave the constant λ a value of 1.89×10^{-5}. Finally, Rutherford was able to repeat the experiment of slow growth, using the emanation diffusing from a paper package of thorium oxide as a steady source of the excited radioactivity, and found that Eq. (4) applied here also, although with the smaller value for the constant λ (Paper 6 in Vol. I).

Perhaps a week after Rutherford had finished the second of his two papers–the one devoted to the excited radioactivity–the transatlantic mails brought him the latest report by the Curies with the description of a transient radioactivity which a strong preparation of radium could induce in neighboring bodies. It must have been a tantalizing document. Their induced radioactivity and his excited radioactivity were possibly similar, but his experiments and theirs differed so widely as to make comparison impossible. In particular, the Curies mentioned nothing which resembled his emanation and asserted positively that their phenomenon did not involve any material deposit but originated rather in a direct transfer of energy from the radium to its surroundings.

As it turned out, an emanation of radium did exist, but its discovery was left for Ernst Dorn of Halle, who could take hints from both Rutherford and the Curies, and who announced it in the spring of 1900.[26]

Actinium and Uranium X, 1899–1900. The Curies were preoccupied with radium, but suspecting the existence of still other radioactive elements they persuaded André Debierne, a former student at the École Municipale, to continue the investigation of the pitchblende residues. There he found in the fall of 1899 a radioactivity in the iron group of his analytical scheme which presently separated with titanium. By the spring of 1900, he was concentrating the radioactive substance with thorium, and since he could distinguish it clearly from both radium and polonium he claimed it as a new element and proposed to name it *actinium.* He speculated that it might prove to be the active material in Rutherford's thorium.[27]

In London, Sir William Crookes was drawn into radioactivity for quite different reasons. He was a distinguished scientist of considerable versatility, the owner and editor of a weekly paper *The Chemical News*, a consulting chemist by profession, and a man of means who was well known for his independent researches in both physics and chemistry. He was attracted by the mysteries of radium, and as a practising chemist chose to extract his own supply from pitchblende. He went about his preparations methodically, arranging for the bulk operations to be performed in a manufacturing plant, and planning the means by which he would assay his growing concentration of

---

[26] E. Rutherford, "A Radio-active Substance emitted from Thorium Compounds," *Phil. Mag.* [5], 1900, *49*: 1–14; "Radioactivity produced in Substances by the Action of Thorium Compounds," *ibid.*, pp. 161–192.

P. Curie and Mme. M.-P. Curie, "Sur la radioactivité provoquée par les rayons de Becquerel," *C. R. Acad. Sci., Paris*, 1899, *129*: 714–716.

E. Dorn, "Über die von radioaktiven Substanzen ausgesandte Emanation," *Abhandlungen der naturforschenden Gesellschaft zu Halle*, 1901, *23*: 15 pp. (separate pagination). Read at the session of June 23, 1900.

[27] A. Debierne, "Sur une nouvelle matière radio-active," *C. R. Acad. Sci., Paris*, 1899, *129*: 593–595; "Sur un nouvel élément radio-actif: l'actinium," *ibid.*, 1900, *130*: 906–908.

radium. He knew that in the early stages he could detect it only by its radioactivity, and as a pioneer photographer and chemist he preferred the photographic plate to the electrometer for registering the intensity of its rays. He was aware of course of the many circumstances in manufacture and processing which could affect the blackening of the photographic image, and he planned to calibrate each plate with the impression of a radioactive standard.

No standard seemed more suitable to him than a pure uranium salt, and with the seasoned distrust of an old-fashioned chemist, he undertook the purification himself. Starting with commercial uranyl nitrate, he performed an ether extraction, followed by a series of fractional crystallizations, and obtained a specimen of undoubted purity which unfortunately produced no blackening of his photographic plates.

Intensive tests assured him that chemical operations in themselves did not destroy radioactivity. In his initial extraction, he had dissolved uranyl nitrate in ether, releasing its water of crystallization to form a separate layer in which some of the uranyl nitrate and the bulk of the impurities would dissolve. This water he had of course discarded. Now he repeated the extraction and found in the water layer the missing radioactivity together with the portion of uranyl nitrate and the impurities that he expected.

There was a suggestion here that the radioactivity belonged to an impurity rather than to the uranium, and after some searching, Crookes found that in fact it did. When he treated a uranyl nitrate solution with ammonium carbonate in excess, the uranium remained in solution as a complex salt, but a thin, fluffy, brown precipitate formed which was obviously aluminum hydroxide colored by the presence of ferric hydroxide. It was also intensely radioactive, and thus contained along with these common elements a substance chemically different from uranium which Crookes proposed to name provisionally "uranium X"* (Paper 7 in Vol. I).[28]

Crooke's announcement was read before the Royal Society in May 1900 and published in *The Chemical News* in June. By this time, Debierne had made an interesting discovery. If he prepared a solution containing actinium and a salt of barium and then precipitated the barium as the sulfate, the precipitated barium proved to be radioactive. As far as he could judge, this radioactivity was a pro-

---

* For historical consistency, the discoverers' names for the radioactive substances are used throughout this essay. Those who wish to identify them by their modern isotopic designations, or by their locations in the genetic chains, are referred to the appendix on p. 253.

[28] Sir W. Crookes, "Radio-activity of Uranium," *Proc. Roy. Soc., Lond.*, 1899–1900, *66*: 409–422; *Chem. News*, 1900, *81*: 253–255, 265–267.

SIR WILLIAM CROOKES (1832-1919)
[From a painting by A. Ludovici. Courtesy National Portrait Gallery, London.]

perty of the barium itself which no chemical operations could remove, and he characterized it then as an induced radioactivity in the Curies' sense of that term, the result of a transfer of energy from the active actinium to the inactive barium during their intimate association in the solution.

Even before Debierne had published, Becquerel borrowed his new technique to attack a problem concerning the radiation from uranium. He had recently observed that some of the rays were deviated by a magnetic field and he wondered whether these deviable rays should be assigned to uranium or to some radioactive material mixed with it. In this case as well the precipitated barium was radioactive, and Becquerel also found that after a series of eighteen successive precipitations from its solution, the activity of one specimen of uranyl chloride was reduced to one-sixth of its initial value.[29]

It seemed a reasonable conclusion then that uranium possessed no radioactivity of its own. What had passed for its radioactivity must belong to a chemically different substance (if indeed it was a specific substance at all), something apparently as rare as polonium and radium.

If the radioactivity of uranium seemed to be growing less tangible, radium was simultaneously becoming more credible. As July was turning to August, Demarçay reported that from the purest of Marie Curie's preparations he had obtained a spectrum of radium containing only the strongest of the barium lines, an indication that the barium was now reduced to a minor constituent of the mixture. Working with a larger sample, amounting to 0.4 grams of mixed radium and barium chlorides, Marie Curie determined an average atomic weight of 174 for the mixture, considerably greater than the value for pure barium of 138.[30]

The Emanations, 1901. The emanations with their transient radioactivities can hardly have seemed as important in 1901 as radium or uranium. Nevertheless there were questions to be answered about their nature and about their relation to the thorium and radium preparations that released them, and these were the questions which Rutherford pursued in his researches of that winter. He had thorium oxide which released emanation as it stood in the laboratory and a

---

[29] A. Debierne, "Sur du barium radio-actif artificiel," *C. R. Acad. Sci., Paris*, 1900, *131*: 333–335.

H. Becquerel, "Note sur le rayonnement de l'uranium," *C. R. Acad. Sci., Paris*, 1900, *130*: 1583–1585, *131*: 137–138.

[30] E. Demarçay, "Sur le spectre du radium," *C. R. Acad. Sci., Paris*, 1900, *131*: 258–259.

Mme. Curie, "Sur le poids atomique du baryum radifère," *ibid.*, pp. 382–384.

commercial specimen of mixed barium and radium chlorides which did not. Both released emanation freely when they were heated, and released it rapidly but in limited quantity when the heating was vigorous. After such heating, the thorium oxide released far less emanation at room temperature than before. All this suggested that the crystals of thorium oxide had occluded a definite amount of emanation, and once this was driven off no more was left to follow. On the other hand, Rutherford found that if he kept the heating moderate, limiting the release of emanation to a lower rate, the total quantity of emanation he could obtain from a thorium oxide specimen was far greater than the burst from an equal amount of the oxide at high temperature. That suggested, rather, that emanation was steadily produced in something like a chemical process.

To learn something of the nature of the stuff, Rutherford conducted a diffusion experiment with radium emanation, whose radioactivity had the relatively long life which was needed. From the rate of diffusion he observed, he concluded that its molecular weight must lie between 40 and 100. (The modern value is 222.) To obtain similar information concerning his other radioactive material, he measured the mobility in electric fields of the positively charged carriers of the excited radioactivity. The results indicated that these carriers were of the size of ordinary gaseous ions, but contemporary opinion was still divided whether ordinary ions were single molecules or molecular clusters.[31]

At some time during the year, Rutherford enlisted the help of Frederick Soddy, a young Oxford graduate who had joined the McGill faculty as Demonstrator in Chemistry only the summer before. Soddy was to carry the chemical side of the investigation, in particular to establish the chemical relationship of the emanation to thorium and to discover what its own chemical nature might be.

The chemistry of thorium was well established. Soddy purified various compounds of it by approved analytical techniques, and found that his final preparations always released the emanation. He concluded thus that the emanation originated from the thorium and not from some intruding impurity.

To investigate the possibility of continuous production, he took some of Rutherford's "de-emanated thoria," samples of thorium oxide

---

[31] E. Rutherford, "Einfluss der Temperatur auf die „Emanation" radioactiver Substanzen," *Phys. Z.*, 1900–01, *2*: 429–431; "Radio-activity," *Nature, Lond.*, 1901, *64*: 157–158; "Transmission of Excited Radioactivity," *Bulletin of the American Physical Society*, 1901, *2*: 37–43; "Übertragung erregter Radioaktivität," *Phys. Z.*, 1902, *3*: 210–214.

E. Rutherford and Miss H. T. Brooks, "The New Gas from Radium," *Transactions of the Royal Society of Canada* [2], 1901, *7*: Sec. III, 21–25.

whose power to release the emanation had been destroyed by vigorous heating. If the thorium produced emanation continuously, then it was possible that chemical treatment might restore the power to release it. If, however, there had been a limited stock which was now exhausted, no restoration could be expected. Soddy found that if he passed the preparation through an aqueous solution and reconstituted the oxide, its emanating power was distinctly increased. If he converted it to the hydroxide, it released emanation even more freely than in its original state, even up to twice the rate after suitable aging.

To determine the chemical properties of the emanation he prepared a series of chemical traps, each designed to form an easily captured compound with one of the known gases. He then sent the emanation through them one after the other, carried along each time by a suitable non-reacting gas. Not one of the traps removed it; it appeared quite as inert chemically as any gas in Ramsay's newly discovered argon family, and might possibly be another member of that group.

It had been part of Soddy's original program to test whether a freely emanating specimen of thorium oxide might eventually show a detectable loss in weight. A simple calculation had showed that to be implausible. In electrolysis, a gram of hydrogen transports a charge of approximately 10^5 coulombs. Rutherford estimated that the quadrant electrometer he used for ionization measurements would respond to 3×10^{-13} coulombs, which would be the charge transported by 3×10^{-18} grams of hydrogen. It was evident then that ionization measurements dealt with quantities of matter far below the capabilities of the balance.

All of these represented Soddy's successes; but when he studied emanating power, the rate at which various preparations of thorium released the emanation, he could find no regularity. High temperatures increased it and low temperatures reduced it, as was to be expected, but it also varied from one compound to another. Thorium oxide released emanation freely, thorium hydroxide very freely, thorium nitrate sparingly in the solid state but copiously from solution. Then he turned to investigate the carbonate, and in dividing a thorium nitrate solution between two precipitates, he committed an error in chemical procedure, and obtained, as he supposed, a thorium carbonate that gave almost no emanation and a hydroxide that released it freely.

The intrusion of the error made this result impossible to repeat, and the confusion which followed drove Soddy back finally to the simple precipitation of thorium hydroxide from a solution of the nitrate. Here he retained both precipitate and filtrate, and found to his pleasure that the residue from the evaporation of the filtrate was strongly radioactive and the hydroxide had lost a good part of its original radioactivity. Moreover this residue, which was substantially free of thorium, released the emanation. Of equal interest, the penetrating power of its rays matched that which Owens had determined for the rays from thorium oxide. Although earlier Soddy had ruled out such a possibility, it now seemed certain that he had managed to isolate a "thorium X," a substance chemically different from thorium but with the radioactive behavior previously attributed to it[32] (Paper 8 in Vol. I).

At this point the calendar intervened to produce a dramatic interruption. It was mid-December. The discovery of thorium X was definite, and although Rutherford and Soddy had still to explore its properties, that exploration might be put off until after Christmas. They closed the laboratory; a few weeks later they returned to find the whole direction of their research altered.

Thorium X as the Sign of Transmutation, 1901–1902. While Rutherford and Soddy studied thorium emanation, Becquerel had gradually become aware of the incongruity of certain well-established facts of experiment. It was easy to remove the radioactivity from uranium. More than a year before he had come close to that feat with his co-precipitations, and Crookes had accomplished it completely. It was also true that during the last five years a great many people had studied uranium, working in different countries with different compounds prepared from different sources. Universally they had found uranium to be radioactive. Then it followed logically either that all commercial refining of uranium must be remarkably ineffective, or that the lost radioactivity must be regenerated after each removal. Neither conclusion was plausible, but at least the second could be tested, since every preparation from his last investigation was still at hand. Becquerel tried them and found his logic vindicated. Every radioactive precipitate of barium sulfate had gone dead and every sample of uranium had regained its original activity.

Uranium X might be a substance as Crookes had supposed, with its own unique set of soluble and insoluble compounds. On the other hand, Becquerel had always regarded radioactivity as a release of stored energy. In that case, what Crookes called uranium X might represent, in the language of Pierre Curie and Debierne, an induction of energy into some non-uranium material in the preparation. Since Becquerel was attracted to both these ideas at once, his

---

[32] E. Rutherford and F. Soddy, "The Radioactivity of Thorium Compounds. I. An Investigation of the Radioactive Emanation," *J. chem. Soc.* 1902, *81*: 321–350.

explanation became a valiant confusion of material and non-material hypotheses[33] (Paper 9 in Vol. I).

Becquerel's report met Rutherford and Soddy on their return to the laboratory, and they needed little time to discover that their preparations had behaved similarly. Their thorium hydroxide had regained its lost radioactivity and their thorium X was dead.

Rutherford knew how to deal with temporary radioactivities, and the partners proceeded to observe in detail just how the thorium X in the filtrate residue lost radioactivity and how the precipitate of thorium hydroxide regained it. For the thorium X, the activity fell off by the geometric decay of Eq. (1); for the thorium hydroxide it rose substantially as prescribed by Eq. (4). In each case, the half-value period was 4 days.

The decay of the thorium X thus followed the same law as the decay of the emanation and of the material which supported the excited radioactivity. As for the hydroxide precipitate, its radioactive growth matched that of Rutherford's paper-package experiments. If the analogy held, that growth must represent the steady addition of thorium X to the precipitate, offset more and more by the exponential decay of the thorium X's radioactivity. Mathematically, at least, the situation was clear. Considered chemically, if thorium X was a substance at all, then it was chemically different from thorium; yet it was apparently from the thorium in the precipitate that the thorium X was derived. This was very like an alchemical transmutation.

It was one thing to draw the conclusion, it was something else to believe it, and quite another thing to convince the rest of the world. Rutherford and Soddy tried other precipitations to establish the chemical properties of thorium X and to assure themselves that they were not dealing with Becquerel's induction of radioactive energy. They carried out repeated separations on the same sample to produce direct evidence of the regrowth of thorium X, and after each waiting interval found in the carefully purified preparation just that replacement of thorium X which Eq. (4) predicted. They tried unsuccessfully to alter the rate at which thorium X was produced from thorium, and took heart from their failure since it set this process apart from ordinary chemical reactions.

While Rutherford and Soddy were accumulating these corroborative details, the Curies published a vigorous attack on Becquerel's recent paper, particularly on his explanation of the regrowth of radioactivity in the uranium preparations. They pointed out that

---

[33] H. Becquerel, "Sur la radioactivité de l'uranium," *C. R. Acad. Sci., Paris*, 1901, *133*: 977–980.

radioactivity was an atomic property, and that Becquerel was consequently advocating an atomic transformation, a phenomenon for which they stated they had seen no evidence in any of their own work. Furthermore they declared that speculation of this nature was premature. At the present stage of knowledge, only general propositions concerning radioactivity might properly be advanced. For their own part they held two to be sufficient as a theoretical basis for research: that radioactivity was an atomic property of matter, and that each radioactive atom should be regarded as a constant source of energy. (Since the Curies had already noticed a slow decrease in the radioactivity of their polonium, they were forced now to declare that it was probably not an element. The public repudiation would embarrass them before the year was out)[34] (Paper 10 in Vol. I).

Faced with this criticism, and with the certain prospect of more from other sources, Rutherford and Soddy drafted the announcement of their new ideas with the greatest care, presenting their experiments first and building up the evidence slowly. Without mentioning the Curies directly, they took pains to meet their arguments, and even, by a slight manipulation, to draw advantage from them. It had been difficult, Rutherford and Soddy pointed out, to accept the radioactive materials as unvarying, and thus as unlimited emitters of energy. With the new discoveries, this was no longer necessary. The radiation came rather from substances of limited period, uranium X and thorium X; the apparent constancy of their radiation arose actually from the steady rate at which those substances were being formed. For this crucial process of formation of thorium X, they used carefully through all the length of their paper the noncommittal description of "chemical change." Only at the very end did they propose explicitly the idea which the whole of it implied, that radioactivity was the external signal of a spontaneous transmutation[35] (Paper 11 in Vol. I).

The Function of the Rays in Transmutation, 1901–1902. It was recognized that the rays of the radioactive materials were not homogeneous. In 1898, Rutherford had distinguished and named two components in the radiation from uranium, the easily absorbed alpha rays and the penetrating beta rays. In 1900, Becquerel and the Curies had found two components again in the case of radium, one penetrating and deviated by a magnetic field, the other easily absorbed and non-penetrating. Whether these bore any relation to

---

[34] P. Curie and Mme. S. Curie, "Sur les corps radioactifs," *C. R. Acad. Sci., Paris*, 1902, *134*: 85–87.

[35] E. Rutherford and F. Soddy, "The Radioactivity of Thorium Compounds. II. The Cause and Nature of Radioactivity," *J. chem. Soc.*, 1902, *81*: 837–860.

Rutherford's uranium rays was an open question, although Becquerel had demonstrated the presence of deviable rays in the uranium radiation. Becquerel had also identified the deviable rays of radium with Thomson's corpuscles or electrons. More recently, a third component of very penetrating, non-deviable rays had been discovered for radium by Paul Villard.[36]

To add something substantial to these scraps of knowledge, Rutherford started A. G. Grier, an electrical engineering student whom he had annexed, on a general investigation of all the radiations in the fall of 1901. In particular, Grier was to study how closely the radiations of uranium, thorium, and radium resembled one another, and to find the relationship between the penetrating and non-penetrating rays. It was generally supposed that the electrons of the penetrating rays made up the primary component of the radiation, and that the non-penetrating component consisted of soft x rays excited by those electrons which were brought to rest within the radioactive material. Grier accumulated a good many observations and found enough general similarity to warrant the extension of Rutherford's terminology from uranium to all the other radioactive substances.

While Grier measured his alpha and beta rays, Rutherford and Soddy were testing their ideas concerning the production and decay of thorium X. The case they were preparing would be stronger if those ideas could be extended to at least one other substance, preferably to uranium X. Unfortunately, the operations which had been effective for Crookes failed completely in Soddy's hands. The radioactivity of his uranium samples was never diminished and his electrometer showed nothing where Crookes had detected uranium X. It was only when in desperation he imitated Crookes precisely, exchanging the electrometer for a photographic plate, that his luck changed.

Then he understood what had happened. The electrometer, as Rutherford had taught him to use it, responded to the heavy ionization of the alpha rays. The alpha rays, however, would be stopped by the same wrappings which screened a photographic plate from light. The photographic exposure could only be produced by beta rays. Then it must follow from his experiments that uranium emitted nothing but alpha rays and uranium X, nothing but betas. Grier had developed special ionization chambers for measuring each component of the radiation in the presence of the other. With them

---

[36] H. Becquerel, "Note sur le rayonnement de l'uranium," *C. R. Acad. Sci,, Paris,* 1900, *130*: 1583–1585.

P. Villard, "Sur le rayonnement du radium," *ibid.*, pp. 1178–1179.

he confirmed Soddy's inferences and then went on to discover that just as with uranium, thorium emitted nothing but alpha rays while thorium X emitted rays of both kinds. Thus alpha and beta rays could be emitted independently. This was a notable discovery, although it negated a good fraction of the work Grier had accomplished.[37]

The independence of the rays was surprising, and quite as surprising was the emission of the alpha radiation by uranium and thorium. So far, Rutherford and Soddy had taken it for granted that the transmutation process preceded the radiation. They supposed that within the transmuted atom a certain quantity of stored energy became available, and that this available energy was then dissipated gradually in radiation, over the minutes or hours or days appropriate to the particular substance. The geometric law of decay prescribed by Eq. (1) could be taken equally well to represent the energy emission of a single atom or the integrated output of the entire sample.

This scheme was now completely overthrown. The alpha rays which belonged to uranium and thorium must be emitted during the transmutation or even perhaps before it. However, this change in detail did not affect their general theory and Rutherford and Soddy were content to leave it unexplored through the minor activities of a quiet summer. Rutherford digressed into a small experimental study of the very penetrating radiation,[38] to which a few months later he gave the name *gamma rays*.[39] He and Soddy began to rewrite their joint papers of the previous winter which had been published in the *Journal of the Chemical Society* [of London] but which would be read by more physicists if they were republished in *The Philosophical Magazine*, the principal British journal of physics. The transformation theory and the evidence to support it made up the first paper; into the second went all their experiments on the properties of the emanation. When this paper was finished, toward the end of the summer, they closed it with a brief and totally irrelevant postscript concerning the emission of the rays.

Here they proposed that the emission of radiation neither pre-

---

[37] E. Rutherford and A. G. Grier, "Magnetische Ablenkbarkeit der Strahlen von radioaktiven Substanzen," *Phys. Z.*, 1901–02, *3*: 385–390; "Deviable rays of Radioactive Substances," *Phil. Mag.* [6], 1902, *4*: 315–330.

F. Soddy, "The Radioactivity of Uranium," *J. chem. Soc.*, 1902, *81*: 860–865.

[38] E. Rutherford, "Sehr durchdringende Strahlen von radioaktiven Substanzen," *Phys. Z.*, 1901–02, *3*: 517–520; "Penetrating Rays from Radio-active Substances," *Nature, Lond.*, 1902, *66*: 318–319.

[39] E. Rutherford, "Die magnetische und elektrische Ablenkung der leicht absorbierbaren Radiumstrahlen," *Phys. Z.*, 1902–03, *4*: 235–240; "The Magnetic and Electric Deviation of the easily absorbed Rays from Radium," *Phil. Mag.* [6], 1903, *5*: 177–187.

PAUL VILLARD (1860-1934)
[From *Mémoires de l'Académie de l'Institut de France* (Paris), Vol. 63, 1941.]

ceded nor followed but accompanied the process of transmutation. In that case, the intensity of the radiation must become an exact measure of the rate at which transmutation was occurring. In the kinetics of chemical reactions, an exponential law in the form of Eq. (1) was the consequence of a monomolecular reaction, one for which the reaction rate was proportional only to the quantity of the single reagent present. Further, they could see that each of the characteristic radioactivities they observed represented a separate, self-dependent transmutation: of thorium into thorium X, of thorium X into the emanation, of the emanation into the substance which gave the excited radioactivity. Indeed that radioactivity which Rutherford at first had called "excited" must mark still another transmutation occurring within the solid deposit. Finally, since radioactivity and transmutation rate were to be considered proportional, the most rapid formation of each new atomic variety would occur where the radiation was strongest, and this explained what Rutherford had previously observed, that the substance which gave the excited radioactivity was always deposited most copiously where the emanation was most radioactive. Although these arguments still had to be tested by a study of the radioactivity of uranium and radium, their successes with thorium made the new version of their theory remarkably promising[40] (Paper 12 in Vol. I).

The beta rays were already known to be corpuscular; the expulsion of a charged particle seemed an apt accompaniment to whatever process accomplished the transmutation of an atom. While the alpha rays were thought to be secondary, they had seemed of little importance. Now they were worth investigation, and before the end of the year Rutherford had demonstrated that they too were corpuscular, carrying positive charges and having a charge-to-mass ratio (in electromagnetic units) of about 6000, an indication that they were not electrons, but rather of the size of atoms. The failure of earlier attempts to deflect them in a magnetic field had indicated not that they were electro-magnetic radiation, but that on the ionic scale their momentum was enormous.[39]

It is appropriate to add here one other remarkable event of the year 1902, the achievement by radium of the status of an element. On July 21, Marie Curie reported to the Académie des Sciences three separate determinations of the atomic weight of radium, giving values of 225.3, 225.8, and 224.0. Such an atomic weight lay properly within the gap between bismuth at 208 and thorium at 232 and entitled radium to the position which its chemical properties indicated, in the

[40] E. Rutherford and F. Soddy, "The Cause and Nature of Radioactivity: Part I," *Phil. Mag.* [6], 1902, *4*, 370–396; "Part II," *ibid.*, pp. 569–585.

row containing uranium and thorium and in the column of the alkaline earths, filling the vacancy below barium.[41]

Extensions to Uranium and Radium, 1902–1903. It was still necessary to discover whether uranium and radium showed any evidence of transmutation. Now that he understood the situation, Soddy made a new separation of uranium X from uranium, and then he and Rutherford began to measure the beta-ray activity of the two preparations. The measurements ran on for five months, and required a certain virtuosity to keep the electrometer reliable over such a period of time, but they did reveal the anticipated behaviour. Uranium X lost its radioactivity exponentially, and the purified uranium showed the exponential regrowth which indicated the steady production of new uranium X. As required, the half-value period of 22 days was the same for both processes.[42]

While the uranium observations went their deliberate way, an opportunity occurred for yet another demonstration of the material nature of the emanations. So far both emanations had remained invisible and imponderable. If they were to be accepted as gases, they must be made to exhibit as many as possible of the ordinary properties of a gas, one of which was to liquefy at a sufficiently low temperature. In October, a machine for liquefying air was installed at McGill as the gift of Sir William McDonald (who had given and equipped the laboratories and endowed professorships in physics, chemistry, mining, and engineering). Rutherford and Soddy made instant use of it. Before its formal dedication, they had demonstrated that both emanations could be chilled out of an air stream.[43]

Along another line of research, in order to incorporate radium into their scheme of radioactivity they followed the decay of the radiation from its emanation, which was exponential with a half-value period of 4 days. Next they dissolved radium chloride in water to free it as far as possible from traces of emanation and let it recrystallize. Rutherford knew from previous observations that the radium chloride crystals would then occlude any new emanation which might form inside them. The exponential rise in radioactivity of these crystals indicated in the now-familiar style that the emanation was indeed being produced there at a steady rate. They repeated Soddy's

---

[41] Mme. Curie, "Sur le poids atomique du radium," *C. R. Acad. Sci., Paris*, 1902, *135*: 161–163.

[42] E. Rutherford and F. Soddy, "The Radioactivity of Uranium," *Phil. Mag.* [6], 1903, *5*: 441–445.

[43] E. Rutherford and F. Soddy, "Note on the Condensation Points of the Thorium and Radium Emanations," *Proceedings of the Chemical Society*, 1902, *18*: 219–220.

A. S. Eve, *Rutherford*, New York: Macmillan, 1939, pp. 88–89.

trapping experiments with the radium emanation, and found that this emanation also had the chemical properties of the noble gases.

In general, the radioactive transformations seemed completely independent of external conditions. By qualitative tests at extremes of temperature, and by the determination of precise decay curves when thorium X was concentrated in a precipitate or dispersed in solution, open to the atmosphere or enclosed in a vacuum, it had been demonstrated that transmutations went on at an unaltered rate. There was one apparent exception. Emanation evolved more freely from preparations of radium and thorium when they were heated or in solution. This was probably a matter of the release of occluded emanation rather than any variation in its production, but Rutherford and Soddy decided to make sure. They took a specimen of radium chloride which had stood undisturbed, with no evolution of emanation over a month. If the production of emanation had continued unaltered within it, then the accumulation over this period should give the factor I_f of Eq. (4). How much had accumulated could be assessed by dissolving the radium chloride in water, capturing all the gases released and sweeping them into an ionization chamber. Next the radium chloride solution could be stoppered and allowed to stand for a definite time. Since they knew the value of λ, Eq. (4) gave all the information necessary to calculate the new accumulation of emanation during this waiting period. However, if the production of emanation was taking place more rapidly in solution than in the solid, the application of Eq. (4) would predict too little for this second trial. In fact, the prediction proved accurate.[44]

Every radioactive substance they had investigated thus fitted into a common scheme. Rutherford and Soddy also found that, in analogy with thorium and uranium, both radium and its emanation emitted nothing but alpha particles. The beta particles appeared only from the active deposit which the radium emanation laid down.

Their October experiment had shown simply that the emanations would condense. To measure the temperature of liquefaction required more elaborate arrangements; it was March (1903) before they were complete. The radium emanation condensed, as they found, at about -150°C. For the thorium emanation the indications were somewhat vague; they could only estimate the condensation temperature as around -120°C. This lack of precision seemed hardly surprising, however, when they calculated that their total amount of liquid had come to no more than 1500 molecules.[45]

---

[44] E. Rutherford and F. Soddy, "A Comparative Study of the Radioactivity of Radium and Thorium," *Phil. Mag.* [6], 1903, *5*: 445–457.

[45] E. Rutherford and F. Soddy, "Condensation of the Radioactive Emanations," *Phil. Mag.* [6], 1903, *5*: 561–576.

The Energy of Radioactivity, 1903. In the spring of 1903, Soddy was offered a position in Sir William Ramsay's laboratory in London. As a final act of partnership, he and Rutherford drew up a summary of their accomplishments and of the present state of the transformation theory. It was a thoroughly confident document, although hardly eleven months had passed since they had set down their first careful arguments for transmutation, and it ended in a remarkable series of calculations.

Moving easily from one plausible estimate to another, they reached at length the prediction that when its succession of transformations had been completed, the total energy released from a gram of radium would be more than 10^8, perhaps as much as 10^{10} gram calories. In contrast, the formation of a gram of water released no more than 4×10^3. The utter incongruity of the magnitudes showed vividly the difference between chemical reactions and the atomic processes of radioactivity.

By a different line of estimation, they predicted the rate of energy release for radium, arriving at a figure of 15,000 gram calories per year. At this rate of spending, given the totals they had just calculated, the life of radium might be measured in no more than thousands of years. Thus the radium contained in a mineral might be far younger than the mineral itself.

Everything done so far suggested to Rutherford and Soddy that except for their radioactivity, all of the radioactive substances behaved in perfectly normal fashion, with quite definite and quite ordinary physical and chemical properties. If an atom of radium could release such a store of internal energy as they had calculated, then it was a fair assumption that the atoms of common elements might contain comparable stores, to be released not spontaneously like radium, but under appropriate conditions. Here, they wrote, one might find an adequate source of energy for the sun[46] (Paper 13 in Vol. I).

Several months before these calculations were begun, Pierre Curie and Albert Laborde had undertaken to measure the energy which radium actually released. They began with a gram of mixed radium and barium chlorides, about one-sixth of which was radium, and compared its temperature with that of a gram of pure barium chloride by means of a thermocouple. The radium-bearing salt was 1.5°C. warmer. They substituted a small coil of wire for the capsule of radium chloride and measured the electrical energy needed to maintain the same temperature. It amounted to 14 gram calories per hour, a value they checked with a Bunsen ice calorimeter. Repeating

---

[46] E. Rutherford and F. Soddy, "Radioactive Change," *Phil. Mag.* [6], 1903, *5*: 576–591.

the measurements with 0.08 grams of pure radium chloride, they reported a provisional value of 100 gram calories per hour for the energy released by a gram of radium[47] (Paper 14 in Vol. I).

At 100 gram calories per hour, a gram of radium would develop nearly 880,000 gram calories per year. By the time Rutherford and Soddy's speculations appeared in print, their 15,000 gram calories appeared no more than a thoroughly conservative underestimate.

Helium from Radium, 1903. Helium has a romantic history. Its existence was postulated by the chemist Edward Frankland and the astronomer Norman Lockyer to account for certain bright lines in the spectrum of the sun. Its discovery on earth followed hard on the discovery of that unexpected inert gas, argon, which helium resembled so closely as to suggest that there might be a whole family of such unknown gases. Ramsay came across it in 1895 as the gas released when the mineral cleveite was dissolved, and Crookes accomplished the spectroscopic identification of Ramsay's terrestrial gas with Frankland and Lockyer's hypothetical solar element.[48]

In the summer of 1902, when Rutherford and Soddy were drafting their second paper on the properties of thorium emanation, they described its occlusion by thorium oxide and mentioned helium as another example of an occluded gas, pointing out that the minerals which occluded helium were all minerals of either thorium or uranium. Thus there was evidently a connection between helium and radioactivity.[39] By the spring of 1903, they were willing to go farther and suggest that helium might be one of the end-products of the radioactive transformations.[46] When they finally parted at the end of that spring term, they agreed that Soddy, as the chemist, should hunt for the probable connection between uranium and radium, while Rutherford, as the physicist, should demonstrate the production of helium by radium.[49]

Between 1895 and 1900, Ramsay had isolated and identified five noble gases: helium, neon, argon, krypton, and xenon. It is small wonder that he wished now to investigate personally the two radioactive emanations which Rutherford and Soddy had assigned to the

---

[47] P. Curie and A. Laborde, "Sur la chaleur dégagée spontanément par les sels de radium," *C. R. Acad. Sci., Paris,* 1903, *136*: 673–675.

[48] M. E. Weeks, *Discovery of the Elements,* 4th Ed. Easton, Pennsylvania: Journal of Chemical Education, 1939, pp. 380–382.

J. N. Lockyer, "The Story of Helium," *Nature, Lond.,* 1895–96, *53*: 319–322, 342–346.

M. W. Travers, "Ramsey and Helium," *ibid.,* 1935, *135*: 619.

W. Ramsay, "Discovery of Helium," *Chem. News,* 1895, *71*: 151.

W. Crookes, "The Spectrum of the Gas from Cleveite," *ibid.,* p. 151.

[49] Sir E. Rutherford, "Early Days in Radio-Activity, "*Journal of the Franklin Institute,* 1924, *198*: 281–289.

same family. In the spring of 1903, he purchased 20 milligrams of pure radium bromide produced by Giesel's firm of Buchler und Compagnie, and proposed to begin, with Soddy's help, by recording the spectrum of its emanation.

By July, his apparatus of interconnected glass tubes and bulbs was ready. Within it he could dissolve his radium bromide crystals in water, and collect and manipulate the occluded gases, purifying them and transferring them to an electric discharge-tube, without opening the apparatus or risking the loss of any part of his material. When the time came, every operation went smoothly, but the spectrum which he saw under the electric discharge was that of carbon dioxide. (One might guess that alpha particles from the emanation had decomposed organic vapors from the grease which sealed the joints.) To salvage what they could, Ramsay and Soddy froze out the carbon dioxide with liquid air. Although this condensed the emanation as well, a trace of gas still remained, and hardly to Soddy's surprise this displayed in their spectroscope the unmistakable yellow line of helium.

Rutherford was on vacation that summer, and was in fact in London on that particular day. When he heard of this event, he promptly turned over to Ramsay and Soddy the 30-milligram specimen of Giesel's radium bromide which he had recently purchased. With the combined output of 50 milligrams, they obtained a clear view of the six principal lines in the visible spectrum of helium[50] (Paper 15 in Vol. I).

Since helium enters into no chemical combinations, there was no process by which the radium bromide could acquire helium from its surroundings. Whatever it contained must have been formed within the solid grains of the specimen after its last crystallization from solution. Here was concrete evidence that one recognized element, helium, could be formed from another, radium. To the casual scientific reader, this would be far more convincing than all of Rutherford and Soddy's mathematical verifications of the consequences of their theory.

For Rutherford the evidence had an additional significance; it reinforced the suspicion he had been entertaining that the helium occluded in minerals might be an accumulation of spent alpha particles.[51]

The Active Deposit and the Rates of Successive Transformations, 1903–1904. Rutherford and Soddy had calculated the energy of radio-

---

[50] W. Ramsay and F. Soddy, "Gases Occluded by Radium Bromide," *Nature, Lond.*, 1903, *68*: 246.

[51] E. Rutherford, "The Amount of Emanation and Helium from Radium," *ibid.*, pp. 366–367.

EDWARD FRANKLAND (1825-1899)
[From *Famous Chemists: The Men and Their Work* by Sir William A. Tilden. New York: E. P. Dutton & Co., 1921.]

SIR JOSEPH NORMAN LOCKYER (1836-1920)
[From *Life and Work of Sir Norman Lockyer* by Thomasine Lockyer, et al. New York: Macmillan, 1928.]

activity as the total energy carried by the alpha particles. They had needed to estimate the masses and velocities of the alpha particles and the rate at which they were expelled, and such estimates of course were uncertain. Nevertheless the discrepancy between their calculation and Curie and Laborde's measured value was disquieting. Possibly there were other modes of energy dissipation which they had ignored.

Consequently, Rutherford's first project in the fall of 1903 was to measure for himself the heat production of a sample whose alpha-particle activity was changing in time. With the help of Howard T. Barnes, who had been initiated in heat measurements by Callendar, he constructed a differential gas thermometer which would compare the temperature of a radioactive specimen with that of its surroundings. As Curie and Laborde had done, they calibrated it by replacing the specimen with an electrically-heated coil.

Next, they heated the new 30-milligram specimen of radium bromide to drive off its occluded emanation, condensed the emanation with liquid air, and sealed the radium and its emanation separately into identical glass capsules. When they inserted these alternately in the gas thermometer, they could watch the heat production of the emanation decrease and that of the radium increase at the same numerical rates as their alpha radiation would have done. Moreover the sum of the two separate rates of heat production remained constant just as the alpha-ray activity of radium remained constant when its emanation was not removed.

These experiments showed that the release of energy varied in exactly the same way as the emission of alpha particles. Since each alpha particle which emerged signalized an atom which had transmuted, the experiments also linked the energy release firmly to the atomic transformations and to no other process.

Strictly speaking, the phenomena were more complicated than the experiment implied. Radium transmuted to the emanation, the emanation to the active deposit, and within the active deposit Rutherford had observed the two further alpha-particle emissions in fairly rapid succession. The gas thermometer was sensitive, but it responded too slowly to follow the changes in the active deposit. To take them into account, Rutherford and Barnes repeated their measurements with resistance thermometers of lower thermal inertia. Once again, the rate of production of heat varied in the same manner as the emission of alpha particles.

Combining what they saw here with earlier ionization measurements, Rutherford and Barnes could describe the processes within the active deposit in these terms: The immediate transformation product of the radium emanation was one of the two alpha-particle emitters, and it transformed itself rapidly with a half-value period of 3 minutes.

Its successor, however, was not the other alpha emitter; the time scale of the process would not permit this. It was necessary to postulate between them an additional substance which gave off no perceptible rays, but was transformed nevertheless with a half-value period of 34 minutes. The third substance in the chain emitted the second alpha particle, and beta and gamma rays as well, and had a half-value period of 28 minutes. What product it formed, they could not say.[52]

This was a situation which Rutherford could describe mathematically. (A complete statement of his mathematical theory is given in Reference 54.) Suppose, to use his notation, that a particular radioactive substance was being produced at a constant rate n_0. Simultaneously, the substance was transmuting to form a succession product with a transmutation rate proportional to the quantity of the substance present. If P stood for this quantity, then $\lambda_1 P$ would give the rate of transmutation, and the net rate of increase of the substance would be

$$\frac{dP}{dt} = n_0 - \lambda_1 P. \tag{5}$$

If Q stood for the quantity of the succession product into which the particular substance transformed itself, then the net increase of this quantity would be

$$\frac{dQ}{dt} = \lambda_1 P - \lambda_2 Q. \tag{6}$$

In the same way, for a third product

$$\frac{dR}{dt} = \lambda_2 Q - \lambda_3 R. \tag{7}$$

If any one of these substances was isolated from its predecessor, a condition which in this mathematical formulation is equivalent to setting $n_0 = 0$ in Eq. (5) or $P = 0$ in Eq. (6), these differential equations gave a solution in the form of Eq. (1),

$$P = P_0 e^{-\lambda_1 t} \tag{8}$$

$$Q = Q_0 e^{-\lambda_2 t}, \tag{9}$$

matching the exponential decrease in activity which Rutherford had

---

[52] E. Rutherford and H. T. Barnes, "Heating Effect of the Radium Emanation," *Nature, Lond.*, 1903, *68*: 622; 1903–04, *69*: 126; *Phil. Mag.* [6], 1904, *7*: 202–219.

E. Rutherford. "Versuch über erregte Radioaktivität," *Phys. Z.*, 1901–02, *3*: 254–257.

E. Rutherford and Miss H. T. Brooks, "Comparison of the Radiations from Radioactive Substances," *Phil. Mag.* [6], 1902, *4*: 1–23.

seen repeatedly. When there was a constant supply of the first substance, Eq. (5) gave another familiar solution in the general form of Eq. (4),

$$P = \frac{n_0}{\lambda_1}(1 - e^{-\lambda_1 t}) \tag{10}$$

if we suppose that initially $P = 0$.

For more general situations, the solution of Eqs. (5), (6), and (7) became more complicated. Rutherford worked them out for certain straightforward cases, and was pleased to discover how neatly they produced the shapes of curves he had already observed in measurements on the active deposit.

There were now too many radioactive substances for casual names: three new ones in the active deposit of radium and at least two in the active deposit of thorium. Taking a hint from Pierre Curie,[53] Rutherford finally adopted a systematic scheme, naming them by the letters of the alphabet in succession. The 3-minute alpha emitter thus became radium A, the rayless intermediate radium B, the final, multi-rayed product radium C. Using these names, we might represent the successive transformations of radium schematically by

$$\mathrm{Ra} \xrightarrow{\alpha} \underset{4\ \text{days}}{\mathrm{Ra\,Em}} \xrightarrow{\alpha} \underset{3\ \text{min}}{\mathrm{Ra\,A}} \xrightarrow{\alpha} \underset{34\ \text{min}}{\mathrm{Ra\,B}} \longrightarrow \underset{28\ \text{min}}{\mathrm{Ra\,C}} \xrightarrow{\alpha,\,\beta,\,\gamma} ?$$

One of the honors at the disposal of the Royal Society is an invitation to a member to deliver the annual Bakerian Lecture. In 1904 the choice fell on Rutherford, within the first year of his election to the Society and hardly two years after the birth of the transformation theory. In May he sailed from Montreal to London, and there described the stages by which the atoms of radium and thorium transmute, presenting the details of his theoretical equations and backing them with his experimental curves. This lecture forms the closing paper of Volume I[54] (Paper 16).

Early Radiochemistry from Polonium to Radium F, 1898–1906. Before Rutherford sailed for London in the spring of 1904, he knew that the chain of radioactive transformations did not end with radium C. Active deposits of some age, recovered for example from tubes which had once held emanation, exhibited a slowly increasing radioactivity. The substances involved showed points of resemblance to polonium, to

---

[53] P. Curie and J. Danne, "Sur la disparition de la radioactivité induite par le radium sur les corps solides," *C. R. Acad. Sci., Paris*, 1903, *136*: 364–366.

[54] E. Rutherford, "The Succession of Changes in Radioactive Bodies," *Philosophical Transactions of the Royal Society, A*, 1905, *204*: 169–219.

radiotellurium, and to radiolead, and this was hardly odd, since these three substances had all been extracted from the minerals of uranium, which also contained radium.

Following the first successes of the Curies, other chemists had been busy, stirring up a disconcerting confusion with their discoveries and investigations. No one quite appreciated that the alpha and beta radiations were different or that they were specific to particular substances. In consequence, as Soddy had demonstrated, the same chemical procedure might lead to rather different results, depending on the absorbing material interposed between the specimen and the detector of radiation. Again, everyone supposed that the radioactive substances were permanent, as ordinary elements are. No one imagined that the chemical nature of a preparation might alter as it stood, and that it mattered, therefore, how much time intervened between successive operations.

The Curies had discovered and named polonium, finding it associated with bismuth in their chemical analysis of pitchblende[14] (see pp. 6–8 and Paper 2 in Vol. II), and Giesel had modelled his procedures closely on theirs, extracting polonium from other minerals by similar methods[21] (see p. 11 and Paper 5 in Vol. II). Yet it was by no means certain that Giesel and the Curies were working with the same material. The Curies' polonium emitted alpha rays (to use Rutherford's nomenclature), and although its radioactivity decreased with the passage of time, it did so rather slowly. Giesel's polonium gave beta rays and had a distinctly short life.[55] The Curies could claim right of discovery but the decreasing radioactivity made them distrust their polonium, and early in 1902, they went so far as to declare it no more than an activated bismuth[34] (see pp. 21–22 and Paper 10 in Vol. I).

Radiolead created difficulties even in the hands of a single investigator. It was discovered and studied between 1900 and 1902 by K. A. Hofmann of the University of Munich, working with a number of different helpers in the laboratory of the Königliche Academie der Wissenschaften. It occurred in a variety of uranium minerals from which it was extracted together with lead, and from this chemical property it had acquired its name. It had a continuing radioactivity which increased over the time required to enclose a specimen in an electric discharge tube, pump out the air, and bombard the specimen with cathode rays. It was photographically active. Finally, its chemical properties were self-contradictory. It was extracted from pitchblende together with lead, but outside of the

---

[55] F. Giesel, "Ueber radioactives Baryum und Polonium," *Ber. dtsch chem. Ges.*, 1900, *33*: 1665–1668; "Ueber radioactive Stoffe," *ibid.*, pp. 3569–3571.

mineral, a repetition of the same extraction procedure removed an inactive lead and left the radioactivity behind.[56]

Radiotellurium arrived two years after radiolead. It was discovered in 1902, extracted with bismuth from the pitchblende residues of Joachimsthal by Willy Marckwald, Professor of Chemistry at the University of Berlin. Nevertheless, it was certainly not bismuth since metallic bismuth displaced it from solution. It had a lasting radioactivity, which differentiated it from Giesel's polonium, and apparently emitted alpha rays since Marckwald found that its radiation, although intense, could be stopped by a sheet of filter paper[57] (Paper 6 in Vol. II). Marckwald was an experienced and resourceful chemist. As his research proceeded, he found that his new substance resembled tellurium, and named it provisionally *radiotellurium* until he should understand its nature better.[58]

Marie Curie was less impressed by Marckwald's chemistry than by the similarities in origin between radiotellurium and her polonium. Early in 1903 she came to the defense, claiming the elementary nature of polonium once more, maintaining that the two substances were the same and publishing in German, the better to reach Marckwald's particular audience[59] (Paper 7 in Vol. II). Marckwald was willing to grant her priority for polonium, but its chemistry as she described it was not that of the radiotellurium he knew. Working always for a comprehensive chemical definition, he presently found means of extracting an inactive tellurium from his preparations, leaving the radioactivity behind[60] (Paper 8 in Vol. II). As for Giesel, his

---

[56] K. A. Hofmann and E. Strauss, "Radioactives Blei und radioactive seltene Erden," *Ber. dtsch chem. Ges.*, 1900, *33*: 3126–3131; "Ueber das radioactive Blei (Vorläufige Mittheilung)," *ibid.*, 1901, *34*: 8–11; "Ueber das radioactive Blei (2. Mittheilung)," *ibid.*, pp. 907–913; "Ueber das radioactive Blei (3. Mittheilung)," *ibid.*, pp. 3033–3039; Ueber radioactive Stoffe," *ibid.*, pp. 3970–3973.

K. A. Hofmann, A. Korn, and E. Strauss, "Ueber die Einwirkung von Kathodenstrahlen auf radioactive Substanzen," *Ber. dtsch chem. Ges.*, 1901, *34*: 407–409.

K. A. Hofmann and V. Wölfl, "Ueber das radioactive Blei," *Ber. dtsch chem. Ges.*, 1902, *35*: 692–694; "Ueber radioactive Stoffe. I. Ueber radioactives Blei," *ibid.*, pp. 1453–1456.

[57] W. Marckwald, "Ueber das radioactive Wismuth (Polonium) [Vorläufige Mittheilung]," *Ber. dtsch chem. Ges.*, 1902, *35*: 2285–2288.

[58] W. Marckwald, "Ueber das radioactive Wismuth (Polonium)," *Verh. dtsch phys. Ges.*, 902, *4*: 252–254; *Chemiker-Zeitung, 1902, 26*: 895–896; *Phys. Z.*, 1902–03, *4*: 51–54; "Ueber den radioactiven Bestandtheil des Wismuths aus Joachimsthaler Pechblende," *Ber. dtsch chem. Ges.*, 1902, *35*: 4239–4240.

[59] Frau Curie, "Über den radioactiven Stoff „Polonium“," *Phys. Z.*, 1902–03, *4*: 234–235.

[60] W. Marckwald, "Ueber den radioactiven Bestandtheil des Wismuths aus Joachimsthaler Pechblende. III," *Ber. dtsch chem. Ges.*, 1903, *36*: 2662–2667.

conviction was steadily growing deeper that what he called polonium was not radiotellurium and probably not even a distinct element, but rather an induction of radioactive energy in any metal which was associated with radium.[61]

Early in 1904, Soddy was moved to comment on the controversy. From the similarity of their radiation, he was inclined to identify radiotellurium with polonium. He remarked in addition that no substance whose radiation was as intense as Marckwald had reported could possibly have a permanent radioactivity. This was a fruitful suggestion, for although Marckwald was convinced of his own chemical accuracy, he set up the experiments to discover whether radiotellurium might lose activity with the passage of time.[62]

Matters had reached this point, when Rutherford, returning to Montreal after the Bakerian lecture, began his serious study of the later products in the active deposit of radium. He had a ready supply in old emanation tubes, from which the deposit could be rinsed with acid into an open dish. That material gave both alpha and beta particles, but the alpha emitter could be separated from the mixture by dipping a stick of bismuth into the solution in acid. This substance plated out on the bismuth, just as radiotellurium would do, and its alpha particles matched in penetrating power those of a specimen of radiotellurium Rutherford had purchased. As the summer progressed, he noticed that the alpha-particle activity of the crude deposit was increasing, an indication that the alpha-emitting substance must follow the beta emitter in the chain of transformations.[54]

Toward the end of the year, Hofmann's laboratory in Munich published a long chemical study of solutions containing radiolead. From such solutions, in particular, they had been able to extract two radioactivities, one of beta rays that died away in a few weeks, one of alpha rays which could still be detected after the passage of a year. After the extraction, the radiolead had spontaneously recovered its lost radioactivity.[63]

A month later came an announcement from Vienna where Stefan Meyer and Egon von Schweidler had been measuring decay periods. They had one specimen of radiotellurium, another of radiolead which had been used for an unsuccessful spectroscopic investigation. Both

---

[61] F. Giesel, "Ueber Radium und radioactive Stoffe, "*Ber. dtsch chem. Ges.* 1902, *35*: 3608–3611; "Ueber Polonium," *ibid.*, 1903, *36*: 728–729; Ueber Polonium und die inducirende Eigenschaft des Radiums," *ibid.*, pp. 2368–2370.

[62] F. Soddy, "Radio-Tellurium," *Nature, Lond.*, 1903–04, *69*: 347, 461–462. W. Marckwald, "Radio-Tellurium," *ibid.*, p. 461.

[63] K. A. Hofmann, L. Gonder, and V. Wölfl, "Über induzierte Radioaktivität," *Ann. Phys., Lpz* [4], 1904, *15*: 615–632.

exhibited a simple exponential decay of radioactivity, both with the same half-value period of 135 days.[64] Not long after, at the end of January, 1895, Marckwald made his report. With eleven months of observation, his assistants had determined the half-value period for radiotellurium at 139.8 days[65] (Paper 9 in Vol. II).

At almost the same time, Rutherford was ready for publication. He could find evidence of three substances after radium C. The first, radium D, had a long life, and went by a rayless transformation into a radium E which emitted beta particles and had a half-value period of about 6 days. Its transformation product, radium F, was the long-period alpha-particle emitter, whose half life Rutherford gave as 150 days, a figure he was later to reduce to 143.

$$\underset{28\text{ min}}{\mathrm{RaC}} \xrightarrow{\alpha,\beta,\gamma} \mathrm{RaD} \longrightarrow \underset{6\text{ days}}{\mathrm{RaE}} \xrightarrow{\beta} \underset{143\text{ days}}{\mathrm{RaF}} \xrightarrow{\alpha} ?$$

Rutherford had no doubt that his radium D was the substance which Hofmann had obtained from pitchblende as radiolead. He was equally sure that polonium and radiotellurium must be the same as his radium F.[66] Gradually, as the radioactivities they were measuring dropped lower and lower, Meyer and von Schweidler came to agree; so did Marckwald, but only so far as radiotellurium was concerned, since he still regarded polonium as a different substance.[67] Nearly a year later, at the beginning of 1906, Marie Curie published a ten-month study of polonium which showed for it the regular exponential decay and a half-value period of 140 days[68] (Paper 10 in Vol. II). This, at last, was evidence which Marckwald would admit to be

---

[64] S. Meyer and E. von Schweidler, "Untersuchungen über radioaktive Substanzen. III. Über zeitliche Änderungen der Aktivität. Vorläufige Mitteilung," *Anz. Akad. Wiss., Wien*, 1904, *41*: 375–377.

[65] W. Marckwald, "Ueber das Radiotellur. IV," *Ber. dtsch chem. Ges.*, 1905, *38*: 591–594.

[66] E. Rutherford, "Slow Transformation Products of Radium," *Nature, Lond.*, 1904–05, *71*: 341–342; *Phil. Mag.* [6], 1905, *10*: 290–306.

[67] S. Meyer and E. R. von Schweidler, "Untersuchungen über radioactive Substanzen. III. Über zeitliche Änderungen der Aktivität," *Anz. Akad. Wiss., Wien*, 1905, *42*: 83; "Untersuchungen über radioaktive Substanzen (III. Mitteilung). Über zeitliche Änderungen der Aktivität," *S. B. Akad. Wiss., Wien* [IIa], 1905, *114*: 387–395.

W. Marckwald, "Radiotellur und Polonium," *Jb. Radioakt.*, 1905, *2*: 133–136.

W. Marckwald, H. Greinacher, and K. Herrman, "Zeitkonstante des Radiotellurs," *ibid.*, pp. 136–139.

[68] Mme. Curie, "Sur la diminution de la radioactivité du polonium avec le temps," *C. R. Acad. Sci., Paris*, 1906, *142*: 273–276.

Frau Sklodowska Curie, "Über die Zeitkonstante des Poloniums," *Phys. Z.*, 1906, *7*: 146–148.

convincing. He capitulated and withdrew the name of radio-tellurium[69] (Paper 11 in Vol. II).

The Thorium Chain of Transformations, 1904–1908. As the polonium controversy showed, the radioactive characteristics of the elements were useful guides to their identity. Marckwald and Marie Curie had worked along parallel paths; the chemical evidence each accumulated would neither confirm nor contradict the other's claims. Given the multiplicity of chemical reactions such a stalemate was always possible, and this one had been broken decisively by a single item of information, knowledge of the half-value period for the decay of radioactivity of the disputed substances. In the next few years, the most secure progress in radiochemistry would be made by those who combined chemical skill with a mastery of the mathematical theory of transformations.

In the fall of 1904 Otto Hahn, a young organic chemist fresh from the University of Marburg, entered Ramsay's laboratory at University College, London. He had crossed the channel primarily to become fluent in English, but he planned to accomplish that task in congenial surroundings, and while engaged in congenial work.[70]

Ramsay had a quantity of a new mineral from Ceylon, from which he expected to obtain radium. The barium fraction had already been separated, and this he turned over to Hahn with instructions to extract the radium from it. Hahn's professional preparation hardly equipped him for the task, but he could at least guide himself by the literature, and he set to work. He did indeed find signs of radium in the crystalline precipitates, but he noticed an even stronger radioactivity in the mother liquor, which should have been impoverished. The source of that radioactivity proved to be a substance with the chemical behavior of thorium. It released, appropriately, an emanation with a half-value period of one minute, and exhibited a radiation far more intense than that of normal thorium. This combination of properties emboldened him to claim it as new and suggested the name he proposed of *radiothorium*[71] (Paper 12 in Vol. II).

Ramsay was thoroughly impressed and exerted his influence to secure Hahn a position in Emil Fischer's laboratory in Berlin. This

---

[69] W. Marckwald, "Über Polonium und Radiotellur," *Phys. Z.*, 1906, *7*: 369–370.

[70] O. Hahn, "Einige Persönliche Erinnerungen aus der Geschichte der natürlichen Radioaktivität," *Die Naturwissenschaften*, 1948, *35*: 67–74.

[71] O. Hahn, "A New Radio-Active Element which Evolves Thorium Emanation," *Proc. Roy. Soc., Lond., A*, 1905, *76*: 115–117; also "Über ein neues, die Emanation des Thoriums gebendes, radioactives Element," *Jb. Radioakt.*, 1905, *2*: 233–266, "Ein neues radioaktives Element, das Thoriumemanation aussendet," *Ber. dtsch chem. Ges.*, 1905, *38*: 3371–3375.

STEFAN MEYER (1872-1949)
[From *Acta physica Austriaca,* Volume 5:1. Vienna, 1951.]

EGON, RITTER VON SCHWEIDLER (1873-1948)
[Photo Courtesy Innsbruck University Library.]

was an excellent post for an able researcher, but Hahn's only present stock in trade was the discovery of radiothorium and he thought it prudent to spend a year in preparation at Rutherford's laboratory in Montreal.[72, 85]

While he was there, mastering the arts of radioactivity, the existence of radiothorium was confirmed most satisfactorily. Giesel extracted something like it from a barium sulfate precipitate made two years before during an analysis of monazite, a thorium mineral, and G. A. Blanc of Rome, as well as Elster and Geitel of Wolfenbüttel, found it in the sediments of radioactive hot springs.[73]

Of greater potential interest were three studies of the general radioactivity of thorium made in the United States: one by Herbert N. McCoy and W. H. Ross at the University of Chicago, one by Bertram B. Boltwood, a consulting chemist in New Haven, the third by H. M. Dadourian at Yale.[74] All three made use of the phenomenon of radioactive equilibrium.

It was a cardinal principle of Rutherford and Soddy's transformation theory that the rate at which any substance transformed into its successor was proportional to the quantity of that substance present. In any chain of transformations, the relative quantities of the various succession products adjusted themselves automatically with the passage of time. A substance present in excess would transform itself rapidly; one in deficiency more slowly. In the end, each substance in the chain would reach an equilibrium amount when as many new atoms were being created from its predecessor as were disappearing by transformation to its successor.

In terms of Rutherford's mathematical theory, the rates of

---

[72] O. Hahn, "Über einige Eigenschaften der α-Strahlen des Radiothoriums, I [and] II," *Phys. Z.*, 1906, *7*: 412–419, 456–462; "On some Properties of the α Rays from Radiothorium," *Phil. Mag.* [6], 1906, *11*: 793–805.

E. Rutherford and O. Hahn, "Mass of the α Particles from Thorium," *Phil. Mag.* [6], 1906, *12*: 371–378.

[73] F. Giesel, "Über die Thoraktivität des Monazits," *Ber. dtsch chem. Ges.*, 1905, *38*: 2334–2336.

G. A. Blanc, "Über die Natur der radioaktiven Elemente, welche in den Sedimenten der Thermalquellen von Echaillon und von Salins-Moutiers (Savoyen) enthalten sind," *Phys. Z.*, 1905, *6*: 703–707.

J. Elster and H. Geitel, "Beiträge zur Kenntnis der Radioaktivität des Thoriums," *Phys. Z.*, 1906, *7*: 445–452.

[74] B. B. Boltwood, "Activity of Thorium Minerals and Salts," *Amer. J. Sci.* [4], 1906, *21*: 415–426.

H. M. Dadourian, "The Radioactivity of Thorium," *Amer. J. Sci.* [4], 1906, *21*: 427–432.

H. N. McCoy and W. H. Ross, "The Relation between the Radioactivity and the Composition of Thorium Compounds," *Amer. J. Sci.* [4], 1906, *21*: 433–443.

increase in Eqs. (5), (6), and (7), dP/dt, dQ/dt, dR/dt, must all be zero. Then one could form a new set of equations,

$$n_0 = \lambda_1 P = \lambda_2 R = \lambda_3 S = \cdot \cdot \cdot \quad . \qquad (11)$$

As Rutherford had originally written Eq. (5), n_0 stood for the constant rate at which the first transformation product was formed. This could now be written $\lambda_0 N$, where N represented the quantity of the primary radioactive material, thorium or uranium. Then Eq. (11) stated that in any chain of radioactive transformations, the quantity of any product present in the chain must be directly proportional to the quantity of the primary substance N and inversely proportional to its own transformation constant λ.

In the state of radioactive equilibrium it would also be true that the number of atoms transforming over any period of time must be the same for every substance in the chain. Then if there were, for example, five successive transformations along the chain, the total radioactivity for the equilibrium state must be five times as intense as the individual radioactivity of any single substance.

In general, a mineral might be expected to be something of a mixture, and one might expect also that the components of the mixture would vary in kind and quantity according to the accidents of formation. If the different substances in a radioactive mineral had entered it independently, then among different specimens of that mineral neither the total radioactivity nor the proportion of any substance present should bear any fixed relation to the abundance of its major constituent. If, on the other hand, all of the radioactive substances present were formed in a single chain of transformations from a single primary substance, then both the total radioactivity and the quantity of any individual product must be proportional to the quantity of primary substance which the mineral contained.

McCoy and Ross chose to measure the total radioactivity of various thorium minerals, as did Boltwood, and to compare its intensity with the thorium content of each as determined by their own analyses. They found a constant ratio for all minerals; in commercial preparations of thorium the total radioactivity was about half as intense. These results indicated that all the radioactive substances in the mineral must be transformation products of its thorium, including by implication the radiothorium. They indicated also that most of the radiothorium had been removed in the manufacture of the commercial preparations. Dadourian chose to assay the amount of emanation which could be obtained from minerals of known composition. He found it proportional to their thorium content, and since Hahn had connected emanation with radiothorium, his conclusions were essentially the same.

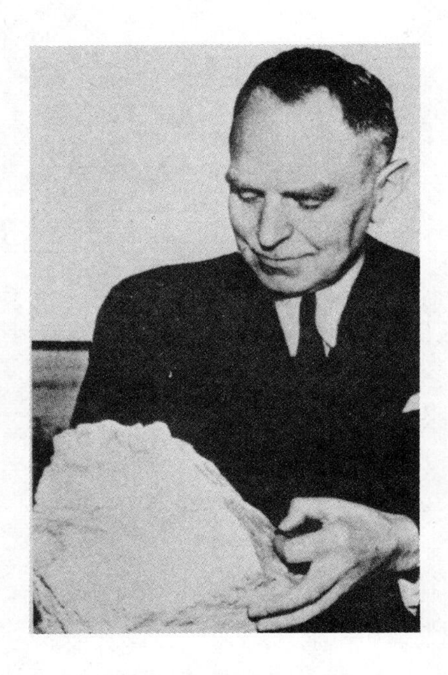

OTTO HAHN (1879-1968)
[From *Discovery of the Elements* by Mary Elvira Weeks. Easton: Journal of Chemical Education, 1956.]

BERTRAM BORDEN BOLTWOOD (1870-1927)
[Photo supplied by The Edgar F. Smith Memorial Collection. University of Pennsylvania.]

This was the state of affairs when Hahn took up his position in Berlin in the fall of 1906. Thorium was in commercial demand for the manufacture of the mantles used in gas illumination, and near Berlin was the chemical firm of O. Knöfler which specialized in high-purity thorium nitrate. The firm was thoroughly willing to co-operate, and gave Hahn access to whatever materials he wished. His immediate goal was to find the point in the manufacturing process at which the thorium and radiothorium separated, but to his surprise they did not seem to do so. Every product that he tested, whether the original mineral, an early crude extract, or the final purified compound, showed substantially the same ratio of radiothorium to thorium.

Thus it was plain that the two were not easy to separate, yet there was evidence for the existence of thorium with a low radiothorium content. The answer which Hahn pieced together from scattered clues was that radiothorium could not be the direct transformation product of thorium. Between them there must occur an intermediate product, apparently rayless and with a relatively long life, which he proposed tentatively to call *mesothorium.* In the refining process mesothorium would separate, leaving thorium and radiothorium together; then in the thorium preparation the radiothorium would decay more rapidly than it could be replaced by the scant quantity of newly generated mesothorium[75] (Paper 13 in Vol. II).

Hahn even had some bits of experimental evidence to back his conjectures, but out of deference to Knöfler, who might find commercial value in mesothorium, he published nothing about its chemical behavior. It was left to Boltwood to suggest that just as radiothorium had proved chemically very like thorium, so mesothorium might resemble thorium X. Indeed, some old preparations in his laboratory indicated as much. On these grounds, Boltwood suggested that Ramsay's barium fraction had removed mesothorium rather than radiothorium from the original mineral. The radiothorium which Hahn had extracted had probably formed there while the material waited on the shelf[76] (Paper 14 in Vol. II).

Meanwhile McCoy and Ross had continued their meticulous investigation of the ratio of radiothorium to thorium in minerals and in chemical preparations of different ages and with different histories. In particular, they worked earnestly to bring about a separation of radiothorium from thorium, but as their tests steadily showed, the two

---

[75] O. Hahn, "Ein neues Zwischenprodukt im Thorium," *Phys. Z.*, 1907, *8*: 277–281; *Ber. dtsch chem. Ges.*, 1907, *40*: 1462–1469.

[76] B. B. Boltwood, "On the Radio-Activity of Thorium Salts," *Amer. J. Sci.*, [4], 1907, *24*: 93–100; "Über die Radioaktivität von Thoriumsalzen," *Phys. Z.*, 1907, *8*: 556–561.

substances maintained a constant ratio of abundance, no matter what sequences of precipitation or crystallization they tried. They concluded that to separate these two by chemical means was "remarkably difficult, if not impossible"[77] (Paper 15 in Vol. II).

During all this time, through 1907 and into 1908, Hahn studied his preparations and watched the slow changes in their radioactivity, attempting to trace the genetic relationships and establish the properties of the thorium products. He found, in spite of all suspicions to the contrary, that thorium itself emitted alpha particles, which he could clearly distinguish from those belonging to radiothorium. He found beta particles coming from his mesothorium specimens, and was able to assign them to a short-lived product which intervened between his original mesothorium and radiothorium. Finally, by elaborate analyses of the curves of growth and decay, he was able to demonstrate that three products and no more lay between thorium and Rutherford and Soddy's thorium X: a mesothorium 1, a mesothorium 2, and his radiothorium[78] (Paper 16 in Vol. II). The successive transformations were as follows:

$$\mathrm{Th} \xrightarrow{\alpha} \underset{5.5\text{ years}}{\mathrm{MsTh\,1}} \longrightarrow \underset{6.20\text{ hours}}{\mathrm{MsTh\,2}} \xrightarrow{\beta} \underset{2\text{ years}}{\mathrm{RaTh}} \xrightarrow{\alpha} \underset{4\text{ days}}{\mathrm{ThX}} \xrightarrow{\alpha} \cdots$$

Ionium, the Predecessor of Radium, 1903–1909. When Soddy left Montreal in the spring of 1903, it was agreed that he would try to grow radium from uranium. The test for radium was both sensitive and certain. After about six weeks, radium was essentially in equilibrium with its emanation. The emanation could be released either by heating the radium or by dissolving it in water, and then blown directly into an ionization chamber, where its alpha particles could produce their maximum effect. Once there, its quantity would indicate the quantity of radium from which it came; once there it could be identified in case of doubt by its half-value period. The experiment Soddy projected was to dissolve a quantity of uranium nitrate in water, free the solution of radium by vigorous purification, seal it up and set it away. Some months later he could open the seal, collect the emanation by bubbling air through the solution, and so assay the expected growth of radium.

---

[77] H. N. McCoy and W. H. Ross, "The Specific Radioactivity of Thorium and the Variation of the Activity with Chemical Treatment and with Time," *J. Amer. chem. Soc.*, 1907, *29*: 1709–1718.

[78] O. Hahn, "Über die Strahlung der Thoriumprodukte," *Ber. dtsch chem. Ges.*, 1907, *40*: 3304–3308; "Zur Nomenklatur der Thoriumzerfallsprodukte," *Phys. Z.*, 1908, *9*: 245, 320; "Ein kurzlebiges Zwischenprodukt zwischen Mesothor und Radiothor," *ibid.*, pp. 246–248; "Über das Mesothorium," *ibid.*, pp. 392–404, 448.

HERBERT NEWBY McCOY (1870-1945)
[From *Discovery of the Elements* by Mary Elvira Weeks. Easton: Journal of Chemical Education, 1956.]

The experiment was easy to project; to perform it in London proved difficult because Ramsay's carelessness in handling his new radium had created a high background of ionizing radiation from the active deposit laid down on the walls of the laboratory by escaped emanation. Although conditions were better in Glasgow, where Soddy moved in the fall of 1904, the results continued to be disappointing. The amount of radium he could detect was less by orders of magnitude than he had expected. When Boltwood tried the same experiment in New Haven with more stringent criteria of initial purity, his results were even more sharply negative. A solution of uranium would not grow radium.[79]

Nevertheless, uranium did turn into radium, as McCoy and Boltwood both proved by equilibrium measurements. McCoy showed that the total alpha-particle activity of uranium minerals stood in a constant proportion to their uranium content, and that therefore every radioactive substance they contained belonged to the transformation chain of uranium. Boltwood assayed the radium more directly in terms of the ionizing power of its emanation and found a constant ratio of emanation activity to uranium content for a wide variety of minerals containing different amounts of uranium, collected in different parts of the world, and obtained from strata of different geological ages.[80] By the summer of 1906, thanks to the loan of a standard solution prepared at McGill from Giesel's pure radium bromide, he could be even more specific. In a mineral in radioactive equilibrium, there would always be 3.8×10^{-7} grams of radium to every gram of uranium.[81]

The failure of radium to appear directly from uranium probably signified a long-lived intermediate product which followed uranium X. Boltwood's suspicions fell on actinium, although without any real justification since he already had hints from his equilibrium studies that actinium was less closely related to uranium than radium was.

---

[79] F. Soddy, "The Life-History of Radium," *Nature, Lond.*, 1904, *70*: 30; "The Origin of Radium," *ibid.*, 1904–05, *71*: 294; "The Production of Radium from Uranium," *Phil. Mag.* [6], 1905, *9*: 768–779.

B. B. Boltwood, "Production of Radium from Uranium," *Amer. J. Sci.* [4], 1905, *20*: 239–244.

[80] H. N. McCoy, "Ueber das Entstehen des Radiums," *Ber. dtsch chem. Ges.*, 1904, *37*: 2641–2656; "Radioactivity as an Atomic Property," *J. Amer. chem. Soc.*, 1905, *27*: 391–403; "The Relation between the radioactivity and the Composition of Uranium Compounds," *Phil. Mag.* [6], 1906, *11*: 176–186.

B. B. Boltwood, "Uranium and Radium," *Nature, Lond.*, 1904, *70*: 80; "On the Ratio of Radium to Uranium in Some Minerals," *Amer. J. Sci.* [4], 1904, *18*: 97–103; "The Origin of Radium," *Phil. Mag.* [6], 1905, *9*: 599–613.

[81] E. Rutherford and B. B. Boltwood, "The Relative Proportion of Radium and Uranium in Radio-Active Minerals," *Amer. J. Sci.* [4], 1905, *20*: 55–57; 1906, *22*: 1–3.

Debierne had abandoned actinium soon after its discovery in 1900; then late in 1902 Giesel had extracted something like it from his own pitchblende ores as a substance which released an air-borne emanation and concentrated with lanthanum rather than thorium. In 1903 Debierne, with reawakened interest, announced that his actinium released an emanation which set up an induced radioactivity in its neighborhood. By February 1904 he had measured the half-value periods of decline of the emanation as 3.9 seconds, and of the induced radioactivity as 40 minutes. Meanwhile, he had steadfastly ignored Giesel's efforts to compare the two substances by an exchange of specimens, until by April 1904 Giesel concluded that the differences between them were substantial and that he was justified in naming his substance *emanium*. In the controversy which followed Debierne insisted on his priority, Giesel on his independence, and in the end, as in the case of polonium, the French name prevailed.[82]

From this beginning, a series of transformation products for actinium was worked out in Rutherford's laboratory. In the fall of 1904 Harriet Brooks obtained decay curves for the active deposit which showed a succession of two substances there, the first of which was rayless. Early in 1905 T. Godlewski identified an actinium X which was soluble in ammonia, was the immediate precursor of the emanation, had a half-value period of 10.2 days, and seemed to play a role quite similar to that of thorium X.[83]

Thus by 1905 there were three substances known to emit alpha particles in the actinium series: actinium X, the emanation, and the second product in the active deposit. In the radium series, there were five. If radium grew from actinium, then in the combined series of transformations roughly one-third of the total alpha-particle activity should be assigned to actinium and its products, whereas in

---

[82] F. Giesel, "Über Radium und radioaktive Stoffe," *Ber. dtsch chem. Ges.*, 1902, *35*: 3608–3611; "Ueber den Emanationskörper aus Pechblende und über Radium," *ibid.*, 1903, *36*: 342–347; "Ueber den Emanationskörper (Emanium)," *ibid.*, 1904, *37*: 1696–1699; "Ueber Emanium," *ibid.*, pp. 3963–3966; "Über Aktinium-Emanium," *Phys. Z.*, 1904, *5*: 822–823.

A. Debierne, "Sur la radio-activité induite provoquée par les sels d'actinium," *C. R. Acad. Sci., Paris*, 1903, *136*: 446–449; "Sur la production de la radioactivité induite par l'actinium," *ibid.*, pp. 671–673; "Sur l'émanation de l'actinium," *ibid.*, 1904, *138*: 411–414; "Sur l'actinium," *ibid.*, *139*: 538–540; "Über das Aktinium," [German translation of the previous paper] *Phys. Z.*, 1904, *5*: 732–734.

[83] Miss H. Brooks, "The Decay of the Excited Radioactivity from Thorium, Radium, and Actinium," *Phil. Mag.* [6], 1904, *8*: 373–384.

T. Godlewski, "A New Radio-Active Product from Actinium," *Nature, Lond.*, 1904–05, *71*: 294–295; "Actinium and its successive Products," *Phil. Mag.* [6], 1905, *10*: 35–45.

fact, as Boltwood had noticed, radium and its successors could account for almost all of the radioactivity of a uranium ore. Nevertheless he proceeded to extract the actinium from a carnotite ore in his possession and set it aside over the spring and summer of 1906. By the end of that time, measurable amounts of radium had appeared.[84]

Meanwhile Hahn had identified a radioactinium, which came between actinium and actinium X, emitted alpha particles, had a half-value period of 19.5 days, and behaved quite generally like radiothorium.[85]

Rutherford was able to confirm the appearance of radium in an actinium preparation of his own, but there were discrepancies which made him wary. By the following spring (1907), he knew for certain that radium did not grow from actinium, and that it did grow in a solution containing Hahn's new radioactinium; but while the radioactinium had been increasing to a peak concentration and dying away, the production of radium had remained steady. The progenitor of radium must be another substance which nevertheless had accompanied radioactinium in his chemical operations.[86]

That was hint enough for Boltwood, who proceeded to dose the solution of a uranium mineral with sodium thiosulfate (the photographer's "hypo") and then decompose the latter with a strong acid. The colloidal sulfur which precipitated entrained the parent of radium, made evident by the growing production from the precipitate of an emanation with the characteristic 4-day half-value period. There were also signs of the thorium emanation with its half-value period of 1 minute, but this was only to be expected since the thiosulfate decomposition was a standard reaction for the extraction of thorium.

As Boltwood continued to investigate, a chemical resemblance to thorium persisted as one of the chief characteristics of his new element. It emitted alpha particles with a distinctively short range of travel through air, and of course it produced radium. There was obviously

---

[84] B. B. Boltwood, "The Production of Radium from Actinium," *Nature, Lond.*, 1906–07, *75*: 54; "Note on the Production of Radium by Actinium," *Amer. J. Sci.* [4], 1906, *22*: 537–538; "Notiz über die Erzeugung von Radium aus Aktinium," *Phys. Z.*, 1906, *7*: 915–916.

[85] O. Hahn, "A New Product of Actinium," *Nature, Lond.*, 1905–06, *73*: 559–560; "Ueber ein neues Produkt des Aktiniums," *Ber. dtsch chem. Ges.*, 1906, *39*: 1605–1607; "Über den Ionisationsbereich der α-Strahlen des Aktiniums," *Phys. Z.*, 1906, *7*: 557–563; "Über das Radioaktinium," *ibid.*, pp. 855–864; "The Ionization Ranges of the α Rays of Actinium," *Phil. Mag.* [6], 1906, *12*: 244–254; "On Radioactinium," *ibid.*, 1907, *13*: 165–180.

[86] E. Rutherford, "Production of Radium from Actinium," *Nature, Lond.*, 1906–07, *75*: 270–271; "The Origin of Radium," *ibid.*, 1907, *76*: 126; "Production and Origin of Radium," *Phil. Mag.* [6], 1907, *14*: 733–749.

no other substance which resembled it, and Boltwood exercised the discoverer's right to name it *ionium*.[87]

During that same summer (of 1907), Hahn had occasion to notice the extraordinary degree to which ionium imitated the chemical behavior of thorium. The Knöfler plant had maintained the custom of setting aside an assay sample from each year's production of rigorously pure thorium nitrate. Hahn discovered that the older of these samples released radium emanation and released it in greater quantity the older they were. As he continued his tests, he could perceive an increase in radium emanation even over the period from August to October. The raw material for the Knöfler thorium nitrate was monazite sand, in which uranium was only a minor constituent. Its ionium content must have been very small indeed. Yet this incredibly minute trace of ionium had remained with the thorium through every stage of a very effective purification.[88]

In January 1908, Marckwald and a student of his named Bruno Keetman published a paper confirming the existence of ionium; their researches had led them to it quite independently of Boltwood, and their description of it matched his in detail, particularly in the chemical resemblance to thorium. They also were able to explain how ionium had remained undiscovered so long. Following the Curies, European investigators had regularly turned to the pitchblende residues of Joachimsthal in their search for new elements, and these residues contained very little ionium. The process by which the uranium was recovered left thorium, and hence ionium, in solution, so both had gone down the valley in the waste water.[89]

To sum up their work, Boltwood and Keetman presently published full reports of all they had accomplished, and these reports are reprinted here as Papers 17 and 18 of Volume II.[90]

The Nature of the Alpha Particles, 1902–1908. The nature of the alpha radiation had posed an interesting question ever since Rutherford and Soddy had noticed its presence in the primary transformations

---

[87] B. B. Boltwood, "The Origin of Radium," *Nature, Lond.*, 1907, *76*: 544–545, 589; "Note on a New Radio-Active Element," *Amer. J. Sci.* [4], 1907, *24*: 370–372; "Mitteilung über ein neues radioaktives Element," *Phys. Z.*, 1907, *8*: 884–886.

[88] O. Hahn, "The Origin of Radium," *Nature, Lond.*, 1907–08, *77*: 30–31; "Über die Muttersubstanz des Radiums," *Ber. dtsch chem. Ges.*, 1907, *40*: 4415–4420.

[89] W. Marckwald and B. Keetman, "Notiz über das Ionium," *Ber. dtsch chem. Ges.*, 1908, *41*: 49–50.

[90] B. B. Boltwood, "On Ionium, a New Radio-Active Element," *Amer. J. Sci.* [4], 1908, *25*: 365–381.

B. Keetman, "Über Ionium," *Jb. Radioakt.*, 1909, *6*: 265–274.

FREDERICK SODDY (1877-1956)

SIR WILLIAM RAMSAY (1852-1916)
[From *Les prix Nobel en 1904*. Stockholm, 1904.]

of thorium and uranium in 1902. Rutherford had discovered then that it was corpuscular, consisting of positively-charged particles moving with a speed of some 2.5×10^9 cm/sec and having a charge-to-mass ratio which he estimated at 6000 in absolute electromagnetic units.[40] Since in those units the charge-to-mass ratio of the hydrogen ions in electrolysis was approximately 10,000, the alpha particles were certainly the size of atoms. A few months later, in May 1903, a second measurement which placed their charge-to-mass ratio at 6400 was reported by Theodor Des Coudres of Würzburg.[91]

In July, six weeks later, Ramsay and Soddy found helium occluded in crystals of radium bromide,[50] and Rutherford was willing to hazard the guess that this helium represented an accumulation of spent alpha particles.[51] He would need the next five years to establish that guess as a certainty.

By the spring of 1904, the necessary experiments were under way. Rutherford wished to measure the deflection of the alpha particles in an electric and in a magnetic field, and to measure in addition the total current which a beam of alpha particles carried. Each of these measurements involved difficulties and each required refinement as it progressed.

In December 1904, for example, W. H. Bragg of Adelaide in Australia published a study of alpha particles. He had found that the alpha particles from any single radioactive substance exhibited a sharply defined range of travel in air, and that those emitted from different substances had different and characteristic ranges.[92] This implied that all the alpha particles from a particular substance were emitted with the same kinetic energy and thus the same velocity. Since the deflection experiments depended upon velocity, Rutherford must take pains to confine them to the alpha particles from only a single substance.

The results of the experiments on the alpha-particle current were capricious until Rutherford discovered that the impacts of the alpha particles dislodged quantities of electrons from anything on which they fell. After that, it was a simple matter to apply a transverse magnetic field strong enough to deflect these electrons back to their starting points. By the spring of 1905, the measurements were complete. If he assumed that each alpha particle carried the elementary ionic or electronic charge, which J. J. Thomson had found to be

---

[91] T. Des Coudres, "Zur elektrostatischen Ablenkbarkeit der Rutherford-strahlen," *Phys. Z.*, 1902–03, *4*: 483–485.

[92] W. H. Bragg, "On the Absorption of α Rays and on the Classification of the α Rays from Radium," *Phil. Mag.* [6], 1904, *8*: 719–725.

W. H. Bragg and R. Kleeman, "On the Ionization Curves of Radium, *ibid.*, pp. 726–738.

3.4×10^{-10} electrostatic units, then Rutherford could say from his own measurement of the current that a gram of radium must emit 6.2×10^{10} alpha particles per second. Since there were reasonable estimates of the number of atoms in a gram of radium, it followed that the fractional part of any radium sample which transmuted itself during a year must be 5.4×10^{-4}, and hence that the half-value period for radium was 1280 years.[93]

By an extension of these calculations, Rutherford was able to make immediate use of the magnetic deflection which he succeeded in measuring for the alpha particles from radium C. That deflection depended upon the velocity of the particles as well as their charge-to-mass ratio, and the velocity could be estimated. From the measurements of Curie and Laborde and his own observations with Barnes, Rutherford could set a figure of 31 calories per hour or 3.6×10^5 ergs per second for the energy production of the radium C in equilibrium with one gram of radium. This must represent the total kinetic energy of the alpha particles from the radium C, and since that substance was present in equilibrium quantity, their number would also be 6.2×10^{10} per second. By combining these figures with the magnetic deflection, Rutherford obtained estimates of 2.6×10^9 cm/sec for the speed, and for the charge-to-mass ratio, 6.5×10^3 electromagnetic units per gram in appropriate agreement with his earlier value.[94]

An electric deflection of the alpha particles was needed for a reliable result. The problem was to develop an electric field strong enough to deflect those particles and yet not so strong as to promote sparking between the electrodes. Rutherford's final solution was to evacuate the deflection chamber and to use broad electrodes set so close together that there was no room for an ion-cascade to develop between them. With this device, by the summer of 1906 he had the positive information he needed. The alpha particles of radium C, which were easy to use and typical of all the rest, were emitted with a speed of 2.06×10^9 cm/sec and had a charge-to-mass ratio of 5070.

Now that he knew this number with some certainty, Rutherford had to consider its implications. It came to about half the value for the hydrogen ion, and thus allowed three possibilities for the identity of the alpha particle. With the ordinary ionic charge, twice the mass of a hydrogen atom was required. The alpha particle might then be a hydrogen molecule or a half-atom of helium. If it carried

---

[93] E. Rutherford, "Charge carried by the α Rays and β Rays of Radium," *Phil. Mag.* [6], 1905, *10*: 193–208.

[94] E. Rutherford, "Some Properties of the α Rays from Radium," *Phil. Mag.* [6], 1905, *10*: 163–176.

SIR WILLIAM HENRY BRAGG (1862-1942)
[From *Les prix Nobel en 1915*. Stockholm, 1916.]

HANS GEIGER (1882-1945)
[From *Zeitschrift für Physik,* Vol. 125, 1948.]

twice the ordinary charge, it might be a helium atom of mass four.[95] No element in the periodic table was appropriate for a triple or quadruple charge.

To be certain, Rutherford needed to know what charge a single alpha particle carried. In a sense, he had half of the answer already since he knew the total charge carried away per second from a gram of radium. If in some way he could count (rather than estimate) the number of alpha particles emitted he would gain the other half.

Three years before, early in 1903, Crookes had discovered that when alpha particles excited fluorescence in willemite, the light they produced was not dispersed in a general glow but concentrated in minute flashes or scintillations.[96] The phenomenon was well-known at McGill;[94,97] it was a reasonable assumption that each scintillation marked the arrival of a single alpha particle. No one knew the efficiency of the process, however. There was no guarantee that each arriving alpha particle produced its spark. For his present purposes, Rutherford could hardly use a counting method which gave only an unknown fraction of the number he wanted.

However, it was distinctly possible that if every ion produced by an alpha particle along the line of its flight could be collected, the total charge would be great enough to register on a sensitive electrometer. This was the line which Rutherford chose to follow, but although he worked at it well into the winter of 1907, the margin of sensitivity remained too low and the counts were never reliable.[98]

That was Rutherford's last winter at McGill. In May 1907, he crossed the Atlantic to become Arthur Schuster's successor as Professor of Physics at the University of Manchester.[99] Here he found Hans Geiger, a young Bavarian recently appointed by Schuster as a research fellow, and promptly enlisted him for the counting project.

---

[95] E. Rutherford, "The Mass and Velocity of the α Particles Expelled from Radium and Actinium," *Phil. Mag.* [6], 1906, *12*: 348–371.

E. Rutherford and O. Hahn, "Mass of the α Particles from Thorium," *ibid.*, pp. 371–378.

[96] Sir W. Crookes, "The Emanations of Radium," *Proc. Roy. Soc.*, 1902–03, *71*: 405–408.

[97] O. Hahn, "On some Properties of the α Rays from Radiothorium," *Phil. Mag.* [6], 1906, *11*: 793–805, *12*: 82–93; "The Ionization Ranges of the α Rays of Actinium," *ibid.*, pp. 244–254.

M. Levin, "On the Absorption of the α Rays from Polonium," *Amer. J. Sci.* [4], 1906, *22*: 8–12; "Über die Absorption der α-strahlen des Poloniums," *Phys. Z.*, 1906, *7*: 519–521 [German version of the previous paper].

[98] Letter of H. A. Bumstead to Rutherford, Feb. 28, 1907, quoted in N. Feather, *Lord Rutherford*, London: Blackie, 1940, p. 118.

[99] See the correspondence reported by A. S. Eve, *Rutherford*, New York: Macmillan, 1939, pp. 144–145, 148, 153–154; and by N. Feather, *Lord Rutherford*, London: Blackie, 1940, p. 112.

He still proposed to collect ions, and since the whole number of ions produced by an alpha particle was not enough, he proposed to develop an ionization chamber in which that number could be multiplied. The basis of the invention was the discovery by J. S. Townsend, a close friend of Rutherford's in his Cambridge days, that in a sufficiently strong electric field, gaseous ions might be accelerated enough in the free path between one molecule and the next to ionize the molecules with which they collided. By such a process, the original number of ions might be increased by hundreds or perhaps thousands of times.

The pattern of ionization chamber which Rutherford and Geiger adopted had first been used by P. J. Kirkby, a young colleague who had worked with Townsend after he had transferred to Oxford.[100] It was cylindrical, its outer wall serving as cathode, with a wire stretched along the axis to form the anode. When an alpha particle passed along it, near and parallel to the wall, the electrons set free would be drawn toward the central anode, moving into an increasingly intense electric field, ionizing by collision, and thus setting free additional electrons to join them in their progress toward the center. With the proper combination of gas pressure in the chamber and potential difference between the electrodes, the more massive positive ions could be held to less than ionization energies as they moved outward. In consequence, there would be only one concentrated cloud of electrons, and after it had reached the anode, all further ionization would cease.

The first model counted rather too freely, but after alteration and re-design, the new counting tube steadied down to reliable performance. Through the spring of 1908, Geiger patiently counted electrometer swings, each marking the arrival of a separate alpha particle. His result was 3.4×10^{10} alpha particles emitted per second per gram of radium, about half as much as Rutherford had estimated three years before.[101] At the same time, Rutherford and Geiger

---

[100] J. S. Townsend, "The Conductivity Produced in Gases by the Motion of Negatively-charged Ions," *Nature, Lond*, 1900, *62*: 340–341, *Phil. Mag.* [6], 1901, *1*: 198–227; "The Conductivity Produced in Gases by the Aid of Ultra-Violet Light," *ibid.*, 1902, *3*: 558–576, 1903, *5*: 389–398.

J. S. Townsend and P. J. Kirkby, "Conductivity Produced in Hydrogen and Carbonic Acid Gas by the Motion of Negatively Charged Ions," *ibid.*, 1901, *1*: 630–642.

P. J. Kirkby, "On the Electrical Conductivity Produced in Air by the Motion of Negative Ions," *ibid.*, 1902, *3*: 212–225.

[101] E. Rutherford, "Recent Advances in Radio-Activity," *Electrician*, 1907–08, *77*: 422–426.

E. Rutherford and H. Geiger, "A Method of Counting the Number of α Particles from Radioactive Matter," *Memoirs and Proceedings of the Manchester*

redetermined the total charge carried in a stream of alpha particles, and this together with their counts gave a value of 9.3×10^{-10} electrostatic units for the charge of each. Since previous measurements had placed the fundamental ionic charge somewhere in the range of 3.1 to 4.06×10^{-10} electrostatic units, they were now sure that each alpha particle carried a double charge and must be an atom of helium.

Since they now knew with some precision the mass of the alpha particle and the value of the basic ionic charge, they felt it worthwhile to calculate new and more reliable values of other atomic constants. The basic ionic charge, the charge of one electron, must be taken as 4.65×10^{-10} electrostatic units. The number of atoms in a gram of hydrogen must be 6.2×10^{23}; consequently the mass of a single hydrogen atom would be 1.61×10^{-24} gram, and the number of molecules of any gas in one cubic centimeter under standard conditions of pressure and temperature would be 2.72×10^{19}. On this basis, they could predict that the volume of emanation in radioactive equilibrium with one gram of radium must be 0.585 cubic millimeters, and the helium production from a gram of radium would be 0.43 cubic millimeters per day. Finally, the half-value period of radium must be 1760 years.[102]

This should have been the end of the project, but Rutherford had recently learned the level of sensitivity of spectroscopic analysis, and he saw that it was within his power to produce the spectrum of an accumulation of spent alpha particles. The actual work was assigned to another of his young men named Thomas Royds. They had a glass tube blown with walls so thin that alpha particles could pass through freely, but sufficiently free of fissures that helium gas did not. This tube was filled with radium emanation and enclosed within a wider tube with stout glass walls to stop the alpha particles. After sufficient time had elapsed, the spectrum of helium began to appear in the gas collected from the intermediate space. As direct evidence of the identity of the alpha particles the experiment was a success, but the waiting time was rather longer than Rutherford had expected. The reason, when they understood it at last, was thoroughly plausible. The alpha particles which could penetrate the glass walls of the inner tube could also penetrate the glass walls of the outer. Most of them, crossing the empty space in which they should have accumulated,

---

Literary and Philosophical Society, 1907–08, *52*: No. 9 (3 pp.); "An Electrical Method of Counting the Number of α-Particles from Radio-Active Substances," *Proc. Roy. Soc. A*, 1908, *81*: 141–161.

[102] E. Rutherford and H. Geiger, "The Charge and Nature of the α-Particle," *Proc. Roy. Soc. A*, 1908, *81*: 162–173.

came to rest only when they were irrevocably imbedded within its enclosing wall of glass.[103]

At the close of 1908, just as these experiments were ending, Rutherford was awarded the Nobel Prize in Chemistry. Rather than rehearse once more the details of the transformation theory, he chose to devote his prize lecture to these latest discoveries, using as his title, "The Chemical Nature of the Alpha Particles from Radioactive Substances."

Mesothorium 1 and Radium, 1910. During 1910, both Marckwald in Berlin and Soddy in Glasgow became professionally involved with mesothorium 1.

Marckwald had been sent a radioactive specimen to assay for its radium content. He had measured the intensity of the gamma rays from radium C in the specimen, then as a check the quantity of emanation which it released, knowing that both would be in equilibrium with the radium. From the gamma rays, he calculated 1% of radium, from the emanation 0.2%. It was a curious discrepancy which hinted at the presence of some other gamma emitter. Marckwald found, in fact, that if he added iron chloride as a carrier and precipitated it with ammonia, the gamma-ray substance was entrained. It had a half-value period of 6 days and could be identified as mesothorium 2. A short-lived substance of this kind could be maintained in a preparation only if it were continually replenished by a long-lived predecessor, and so the preparation must contain mesothorium 1. When Marckwald attempted to extract this, however, he found that whatever procedures removed the mesothorium 1 would remove the radium as well. Consequently from a mixed ore of uranium and thorium, one might expect mesothorium 1 in any preparation containing radium.

Marckwald was aware of the close chemical resemblance exhibited by thorium, radiothorium, and ionium, and he knew that Keetman's recent work had added uranium X to the set.[90] In this new discovery, he saw the beginnings of a second group of chemically similar elements[104] (Paper 19 in Vol. II).

By now, Soddy had spent seven years in fruitless efforts to establish the chain of transformations between uranium and radium. His uranium solutions had grown no radium over years of watching; more recently, when he had tried to attack the problem through uranium

---

[103] E. Rutherford and T. Royds, "The Nature of the α Particle," *Memoirs and Proceedings of the Manchester Literary and Philosophical Society*, 1908–09, *53*: No. 1 (3 pp.); "The Nature of the α Particle from Radioactive Substances," *Phil. Mag.* [6], 1909, *17*: 281–286.

[104] W. Marckwald, "Zur Kenntnis des Mesothoriums," *Ber. dtsch chem. Ges.*, 1910, *43*: 3420–3422.

X, he had been frustrated by his failure to discover any chemical reactions which that substance would follow reliably.[105]

In 1910, he decided to acquire some mesothorium, and since none had yet been put on the market, he purchased some thorianite (which contained uranium as well as thorium) and began his own process of extraction. Taking the hint from Boltwood's comments on Hahn's early work, he added barium as a carrier and precipitated barium sulfate. As he anticipated, that precipitate contained the mesothorium, and as he had not anticipated, retained it tenaciously. No procedures less powerful than those used to separate radium would remove it. When at last he had removed the radium and mesothorium together, no further operations that he could devise would alter their relative proportions in the mixture, or for that matter the relative proportions of the mesothorium and thorium X which was also present.

This was not greatly different from McCoy and Ross's results with thorium and radiothorium, nor were Soddy's operations more powerful or thoroughgoing than those they had carried out. Yet where they and all their successors had cautiously asserted that these substances, because of their great similarity, would be "remarkably difficult" to separate, Soddy declared boldly for "impossible." Radium, mesothorium 1, and thorium X, he said, were not merely chemically similar, but chemically identical.

He was perfectly aware that this claim violated an accepted principle of chemistry. If these three were identical, then they were not three different (although closely related) substances but one, a single element with a unique set of chemical properties. The final, stringent test for the identity of an element was that every portion of it exhibited the same atomic weight. As Soddy took pains to point out, these three radioactive materials had three different atomic weights. That

---

[105] F. Soddy, "The Origin of Radium," *Nature, Lond.*, 1907, *76*: 150; "The Relation between Uranium and Radium. III," *Phil. Mag.* [6], 1908, *16*: 632–638; "The Relation between Uranium and Radium. IV," *ibid.*, 1909, *18*: 846–858; "The Rays and Product of Uranium X," *ibid.*, pp. 858–865; "The Relation between Uranium and Radium V," *ibid.*, 1910, *20*: 340–342; "The Rays and Product of Uranium X." *ibid.*, pp. 342–345; "Essais pour évaluer la période de l'ionium," *Le Radium, Paris*, 1910, *7*: 295–300.

F. Soddy and T. D. Mackenzie, "The Relations between Uranium and Radium," *Phil. Mag.* [6], 1907, *14*: 272–295.

F. Soddy and A. S. Russell, "The γ rays of Uranium and Radium," *Phil. Mag.* [6], 1909, *18*: 620–649; "The Decay Constant of Uranium X," *ibid.*, 1910, *19*: 847–851.

F. Soddy and R. Pirret, "The Ratio between Uranium and Radium in Minerals," *Phil. Mag.* [6], 1910, *20*: 345–349.

R. Pirret and F. Soddy, "The Ratio between Uranium and Radium in Minerals. II," *Phil. Mag.* [6], 1911, *21*: 652–658.

of radium had been directly determined as 226.5. Mesothorium 1 was formed from thorium by the expulsion of a single alpha particle. Its atomic weight then must be 232.4 (for thorium) less 4.0 (for the helium atom which constituted the alpha particle) or 228.4. Between mesothorium 1 and thorium X another alpha particle was lost, so the atomic weight of thorium X would be 224.4.

Furthermore, he was willing to make the same claims for two other groups which showed similar differences in atomic weights. He would take as a single element the triad of thorium (232.4), ionium (230.5), and radiothorium (228.4). (On the basis of his own experience, he found it difficult to accept Keetman's claim that uranium X belonged here also.) Finally, he would take as a single element radium D or radiolead (210.4) and ordinary lead (207.1), whose chemical resemblances had been demonstrated at the Curie laboratory in Paris by B. Szilard and H. Herchfinkel.

In fact, he saw no reason to restrict these speculations to the radioactive elements. Among the ordinary elements, he pointed out, there might well be some which were actually mixtures in constant proportions of chemically non-separable substances with atomic weights which also differed by whole numbers. The existence of such mixtures might account for the irregularities in the actual distribution of atomic weights[106] (Paper 20 in Vol. II).

In 1902, Soddy had assisted in destroying the doctrine of the immutability of atoms. Now he was also denying their uniformity.

Systematic Radiochemistry, 1911–1913. Soddy had reached his conclusions with a minimum of experimental support. To buttress them more firmly and to extend them where he could, he spent the better part of 1911 searching the literature for everything then known about radiochemistry, which he summarized and published in a small brown book, *The Chemistry of the Radio-Elements.*

As he assembled his information, he noticed that all the radioactive elements whose chemistry was well understood had even valences and stood in the even-numbered columns of the periodic table. This was true for his tetravalent thorium group in the fourth column and his divalent radium group in the second. It was true for the non-valent emanations in the zero column. It was true for polonium, in the sixth column, below tellurium, where Marckwald had placed it; and if polonium turned into inactive lead, then lead fell properly for the scheme in the fourth column.

---

[106] F. Soddy, "The Chemistry of Mesothorium," *J. Chem. Soc.*, 1911, *99*: 72–83.

B. Szilard, "Étude sur le radioplomb," *C. R. Acad. Sci., Paris*, 1908, *146*: 116–118; "Étude sur le radioplomb," *Le Radium*, 1908, *5*: 1–5.

H. Herchfinkel, "Sur le radioplomb," *Le Radium*, 1910, *7*: 198–200.

KASIMIR FAJANS (b. 1887)
[Photo from Cornell University Archives.]

SIR ALEXANDER FLECK (1889-1968)
[From *Discovery of the Elements* by Mary Elvira Weeks. Easton: Journal of Chemical Education, 1956.]

In the alpha-particle transformations among these elements, such as from thorium to mesothorium 1, or from thorium X to the emanation, the changing element moved from one even-numbered column to the next, skipping over the odd-numbered column between. These particular transformations moved the element down the periodic table, but upward motions occurred also. Radiothorium in the fourth column followed mesothorium 1 in the second, and polonium in the sixth came after radiolead in the fourth[107] (Paper 21 in Vol. II).

There were also radioactive substances whose chemistry was doubtful or unknown, and for some of these Soddy was now in a position to make predictions. Early in 1912, he set up an investigation of a few of them, enlisting for the task a promising young chemist named Alexander Fleck who had taken his B.Sc. at Glasgow only the spring before. Fleck had a new technique to try, a direct application of Soddy's ideas. If it seemed likely that a radioactive substance had identical chemistry with some other, he would mix the two and try to alter the proportions of the mixture. If all his precipitations and crystallizations failed to do so, they must be the same. By this method, he convinced Soddy that uranium X did have the chemistry of thorium, and that so also did radioactinium. By this method also, he demonstrated that thorium B had the chemistry of lead; and he was able to report these three discoveries to the British Association for the Advancement of Science, meeting that September in Dundee.[108]

The time was fully ripe now to fit the radioactive elements into the periodic table. If Soddy's hypothesis of chemical identity was taken seriously one perplexing problem had been solved, that of crowding a large number of substances into a small number of places. The thorium elements, the radium elements, the rare gas elements, radium D and radium F were already placed. In 1913, schemes for disposing of the others came with a rush.

The first to be published was by A. S. Russell, who had studied under Soddy and was now at Cambridge.[109] It was best where it followed Soddy's ideas most closely, and in a few weeks it was superseded by a more closely reasoned system.

This was proposed by Kasimir Fajans, a physical chemist from Warsaw who had taken his doctorate at Heidelberg. From there he had gone to Zürich, then to Rutherford's laboratory at Manchester, and recently he had been admitted as a *Dozent* or lecturer at the

---

[107] F. Soddy, *The Chemistry of the Radio-Elements*, London: Longmans, 1911.

[108] A. Fleck, "The Chemical Nature of Uranium X, Radioactinium, and Thorium B," *Chem. News*, 1912, *106*: 128.

[109] A. S. Russell, "The Periodic System and the Radio-Elements," *Chem. News*, 1912, *107*: 49–52.

THE PERIODIC TABLE OF THE ELEMENTS IN 1912

[From Walther Nernst *Theoretische Chemie*, 7th ed. Stuttgart: Enke. 1913.]

0	I	II	III	IV	V	VI	VII	VIII
He 3.99	Li 6.94	Be 9.1	B 11.0	C 12.00	N 14.0	O 16.0	F 19.0	
Ne 20.2	Na 23.00	Mg 24.32	Al 27.1	Si 28.3	P 31.0	S 32.07	Cl 35.46	
A 39.9	K 39.10	Ca 40.1	Sc 44.1	Ti 48.1	V 51.0	Cr 52.0	Mn 54.9	Fe 55.8 Co 59.0 Ni 58.7
	Cu 63.6	Zn 65.4	Ga 69.9	Ge 72.5	As 75.0	Se 79.2	Br 79.92	
Kr 82.9	Rb 85.5	Sr 87.6	Y 89.0	Zr 90.6	Nb 94	Mo 96.0	?	Ru 101.7 Rh 102.9 Pd 106.7
	Ag 107.88	Cd 112.4	In 115	Sn 119.0	Sb 120.2	Te 127.5	I 126.9	
X 130	Cs 133	Ba 137.4	La 139.0	Ce 140.25	—	—	—	
	—	—	Yb 172.0	—	Ta 181.5	W 184.0	—	Os 191 Ir 193.1 Pt 195.2
	Au 197.2	Hg 200.6	Tl 204.0	Pb 207.1	Bi 208.0	—	—	
	—	Ra 226	—	Th 232.4	—	U 238.5		

ERNEST MARSDEN (b. 1889)
[Photo by M. D. King, Victoria University, Wellington.]

Technische Hochschule in Karlsruhe. Fajans found his organizing principle in electrochemistry. Arguing from the behavior of elements whose chemistry was known, he was able to deduce two rules: that the emission of a beta particle produced a new element which was more electronegative than its predecessor (or electrochemically nobler, as Fajans preferred to phrase it), and that the emission of an alpha particle produced one which was more electropositive. An element produced by the emission of a beta particle must lie, then, to the right of its parent in the periodic table. Fajans decided that it must occupy the next place on the right. An element produced by the emission of an alpha particle must lie to the left, in the second place to the left as Soddy had already shown. A number of transformations were accepted as rayless, and these Fajans grouped with the beta-particle emissions.

With these rules and these principles, Fajans had little trouble fitting all of the elements descended from uranium, thorium, and actinium into the periodic table. At only one point was there any serious difficulty. Radium D belonged with lead in the fourth column, and working up from radium D, Fajans placed radium A in the sixth column near the extreme electronegative edge of the table. Radium A was derived from radium by the successive emissions of two alpha particles. Radium belonged in the second column, near the electropositive edge of the table; by Fajans' rules, radium A should be distinctly more electropositive than radium. To save the situation, Fajans invented a radium X, produced by a beta-particle emission from radium and therefore electronegative, to intervene between radium and the emanation[110] (Papers 22 and 23 in Vol. II).

Shortly after Fajans' paper appeared, Soddy published his own scheme for the periodic table, borrowing details from Russell and Fajans as they had borrowed from him, and tacitly correcting their errors as he did so. From Fajans in particular, he took the rules for locating transformation products, but without the electrochemical principles so that he had no need of the hypothetical radium X. In his arrangement, the emission of an alpha particle by radium shifted the position of the new element two places down the periodic table to the rare-gas column where the emanation belonged. The emission of another shifted it two places still farther, to the position of polonium in the sixth column where radium A probably belonged.

Of the three men, Soddy realized most clearly the implications of chemical identity, what it meant to group such different substances

---

[110] K. Fajans, "Über eine Beziehung zwischen der Art einer radioaktiven Umwandlung und dem electrochemischen Verhalten der betreffenden Radioelemente," *Phys. Z.*, 1913, *14*: 131–136; "Die Stellung der Radioelemente im periodischen System," *ibid.*, pp. 136–142.

as thorium, radiothorium, radioactinium, uranium X, and ionium as varieties of a single element. There had been a controversy over the lifetime of ionium, which radioactive evidence would place at tens of thousands of years. However, two spectroscopic examinations of thorium preparations with a strong radioactivity of ionum had shown the spectrum of nothing but thorium, and this had suggested a small concentration of very active and hence short-lived ionium. Now, as Soddy saw, the discrepancy would vanish if one supposed that thorium and ionium were really identical, even to the point of having the same spectrum[111] (Paper 24 in Vol. II).

When Fleck's full report was published, it served chiefly to confirm what Fajans and Soddy had already done. It confirmed, however, by detailed chemical operations the conclusions they had reached from radioactive premises[112] (Paper 25 in Vol. II). His investigation was followed by others using the same technique of mixture and attempted separation. In every case, the chemical properties were found to be those which Fajans and Soddy had predicted.[113]

Soddy had believed from the start that his ideas about atomic varieties need not be restricted to the radioactive elements, and he saw the value in having a name for these atoms which differed and still were the same. Late in 1913, he casually introduced one. He proposed the name *isotopes* to designate varieties of atoms which were chemically identical but which differed from one another in atomic weight and sometimes also in radioactivity[114] (Paper 26 in Vol. II).

The Utility of the Nuclear Atom, 1911–1913. What Fajans and Soddy had done was supported by observation of both the radioactive and the chemical behavior of the elements. Within a few months, it received additional support from a new theory of atomic structure.

Two years earlier, Rutherford had devised an atomic model to account for the scattering of alpha particles by metal foils. He proposed that within the boundaries of the atom as established by the kinetic theory of gases, there would be nothing to influence the motion of the alpha particle except a single, very tiny, massive, charged scattering-center; and on the basis of this assumption, he derived a

---

[111] F. Soddy, "The Radio-Elements and the Periodic Law," *Chem. News*, 1913, *107*: 97–99.

[112] A. Fleck, "The Chemical Nature of some Radioactive Disintegration Products," *J. Chem. Soc.*, 1913, *103*: 381–399.

[113] K. Fajans and P. Beer, "Über die chemische Natur einiger kurzlebiger Radioelemente," *Naturwissenschaften*, 1913, *1*: 338–339.

K. Fajans and O. Göhring, "Über die komplexe Natur des Ur X," *ibid.*, p. 339.

W. Metzener, "Zur Kenntnis der chemischen Eigenschaften von Thorium C und Thorium D," *Ber. dtsch chem. Ges.*, 1913, *46*: 979–986.

[114] F. Soddy, "Intra-atomic Charge," *Nature, Lond.*, 1913–14, *92*: 399–400.

NIELS BOHR (1885-1962)
[Photo by Herdis and Herm. Jacobsen, Copenhagen.]

FRANCIS WILLIAM ASTON (1877-1945)
[From *Les prix Nobel in 1921-22*. Stockholm, 1923.]

detailed and quantitative prediction for the distribution of the scattered particles.[115] The prediction was the same whether the scattering center should be assigned a positive or a negative charge. By the summer of 1912, however, Rutherford had tacitly assumed that the charge was positive and was beginning to refer to the scattering center as the "nucleus" of the atom.[116] By the spring of 1913, his predictions of scattering distribution were elaborately confirmed by the experiments of Hans Geiger and Ernest Marsden.[117]

During the spring of 1912, Rutherford had entertained as a four-month visitor to his laboratory a young Dane named Niels Bohr, who had completed his doctorate only the year before. Bohr then returned to Copenhagen to elaborate Rutherford's atom model so that it might account for phenomena other than alpha-particle scattering. In the summer of 1913, his elaborations appeared in a series of three papers.[118]

It was understood that the atom contained electrons in sufficient number to make it electrically neutral. Bohr's concern was to dispose them in stable arrangements that would establish the discrete energy states which a quantum theory of spectra required. The particular details of his solution to the problem are not our present concern.

The nuclear model provided an enormous separation between the outer electrons of an atom and its nucleus. This suggested that chemical behavior, which was notably sensitive to the environment of an atom, would be governed by the arrangment of the outer electrons, and that radioactive behavior would be governed only by the structure of the nucleus. Thus, provided that the outer electrons maintained the same pattern around different nuclei, there might be atoms with identical chemistry which differed in radioactivity and atomic weight.

The arrangement of the outer electrons should depend primarily on their number. Bohr assumed that this number was equal simply to the ordinal number of the place the atom occupied in the periodic table, giving credit for the idea to an amateur physicist named van den Broek. This number must also give the positive charge on the nucleus, measured in electronic units. In consequence, what Soddy was soon to call isotopes would be atoms with equal nuclear charges but different masses, and thus might easily differ in radioactivity.

---

[115] E. Rutherford, "The Scattering of α and β Particles by Matter and the Structure of the Atom," *Phil. Mag.* [6], 1911, *21*: 669–688.

[116] E. Rutherford, "The Origin of the β and γ rays from Radioactive Substances," *Phil. Mag.* [6], 1912, *24*: 453–462.

[117] H. Geiger and E. Marsden, "The Laws of Deflexion of α Particles through Large Angles," *Phil. Mag.* [6], 1913, *25*: 604–623.

[118] N. Bohr, "On the Constitution of Atoms and Molecules," *Phil. Mag.* [6], 1913, *26*: 1–25, 476–502, 857–875.

The alpha and beta particles emitted when an atom transformed itself undoubtedly came from the nucleus, and by their separation from it altered the charge it carried. Thus the expulsion of one of these charged particles could accomplish the chemical change in the atom, as Rutherford and Soddy had guessed in 1902. What was more, the loss of an alpha particle lowered the positive nuclear charge by two units, and the loss of a beta particle raised it by one. On Bohr's assumption, these changes were just sufficient to establish the rules which Soddy and Fajans had given for locating the transformed atom in the periodic table.

Conclusion. The isotopes that Soddy had recognized could be distinguished from one another by their radiation and by the past and future histories of the atoms which constituted them. He had imagined others which could be distinguished only by the masses of their atoms, and which could be identified only if they could be sorted by atomic mass. In 1913, the first steps were taken to accomplish that task. Thomson at Cambridge had discovered how to sort the positive ions in an electric discharge tube, and in that year he found that when neon was present, he obtained two new varieties with atomic weights of 20 and 22. They might possibly represent isotopes, and F. W. Aston of the same laboratory, who believed that they did, set up diffusion experiments from which he recovered two small samples of neon gas with a genuine difference in density of about one-half per cent.[119]

It is characteristic of history that it neither begins nor ends. Every epoch grows from preceding events and shapes the pattern for the future. Yet though history is endless, histories must be limited. In 1913, there were still unsolved problems of radioactivity and the concept of isotopes had only begun to be fruitful. Nevertheless, this is an appropriate place to stop. The penetrating rays which had been pure mystery to Becquerel had now been analyzed and identified, the source of their energy localized as the inherent energy of nuclear constitution. The varieties of changing atoms to whose discovery the rays had led had been studied and classified, and their behavior was now predictable. Much remained to be done after 1913, but there is nevertheless a genuine unity in what had been achieved by that time.

---

[119] J. J. Thomson, "Further Applications of Positive Rays to the Study of Chemical Problems," *Proceedings of the Cambridge Philosophical Society*, 1912–14, *17*: 201 (Session of 27 Jan. 1913): "Some Further Application of the Method of Positive Rays," *Nature, Lond.*, 1913, *91*: 333–337; "Positive Rays of Electricity," *ibid.*, p. 362; "Physics at the British Association," *Nature, Lond.*, 1913–14, *92*: 304–309; see also *Engineering*, 1913, *96*: 423.

F. W. Aston, *Isotopes*, New York: Longmans, Green, 1923, chap. III–IV, pp. 22–42.

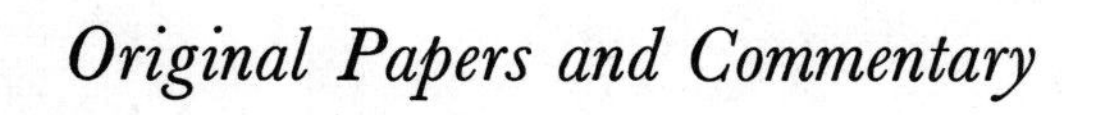

Original Papers and Commentary

I. The Curies: Polonium and Radium

[In 1897, several varieties of rays were known to physics, none of them well understood. By far the most interesting were the x rays because of their spectacular powers of penetration.[1] There were also the cathode rays, which were somehow involved in the production of x rays and which were now being claimed to be corpuscular, notably by J. J. Thomson of Cambridge.[2] There were the Lenard rays, which might simply be cathode rays passed out of their generating tube through a thin, metallic window.[3] Finally there were Becquerel's "uranic rays," which appeared to be emitted spontaneously by the atoms of uranium.[4]

---

[1] O. Glasser, *Wilhelm Conrad Röntgen and the Early History of the Roentgen Rays*, Springfield, Illinois: Thomas, 1934. There is an extensive bibliography of books, pamphlets and research papers on x rays published during 1896 on pp. 422–479.

[2] J. Perrin, "Nouvelles propriétés des rayons cathodiques," *C. R. Acad. Sci., Paris*, 1895, *121*: 1130–1134.

E. Wiechert, "I. Ueber das Wesen der Elektricität," *Schriften der physikalisch–ökonomischen Gesellschaft zu Königsberg in Pr.*, 1897, *Berichte*, [3]–[12]; "II. Experimentelles über die Kathodenstrahlen," *ibid.*, [12]–[16].

J. J. Thomson, "Cathode Rays," *The Electrician*, 1897, *39*: 104–109; *Phil. Mag.* [5], 1897, *44*: 293–316.

W. Kaufmann, "Die magnetische Ablenkbarkeit der Kathodenstrahlen und ihre Abhängigkeit vom Entladungspotential," *Ann. Phys., Lpz* [3], 1897, *61*: 544–552.

[3] P. Lenard, "Ueber Kathodenstrahlen in Gasen von atmosphärischem Druck und im äussersten Vacuum," *Sitzungsberichte der Akademie der Wissenschaften zu Berlin*, 1893, pp. 3–7; *Ann. Phys., Lpz* [3], 1894, *51*: 225–267; "Ueber die magnetische Ablenkung der Kathodenstrahlen," *ibid.*, *52*: 23–33; "Ueber die Absorption der Kathodenstrahlen," *ibid.*, 1895, *56*: 255–275; Ueber die elektrische Wirkung von Kathodenstrahlen auf Atmosphärische Luft," *ibid.*, 1897, *63*: 253–260.

[4] H. Becquerel, "Sur les radiations émises par phosphorescence," *C. R. Acad. Sci., Paris*, 1896, *122*: 420–421 (Reprinted as Paper 1 in *The Discovery of Radioactivity and Transmutation*, ed. A. Romer, in this series, to be abbreviated as Vol. I.); "Sur les radiations invisibles émises par les corps phosphorescents," *ibid.*, pp. 501–503 (Paper 2, Vol. I); "Sur quelques propriétés nouvelles des radiations invisibles émises par divers corps phosphorescents," *ibid.*, pp. 559–564; "Sur les radiations invisibles émises par les sels d'uranium," *ibid.*, pp. 689–694 (Paper 3, Vol. I); "Sur les propriétés différentes des radiations invisibles émises par les sels d'uranium, et du rayonnement de la paroi anticathodique d'un tube de Crookes," *ibid.*, pp. 762–767; "Émission de radiations nouvelles

It was the rays from uranium that Marie Curie (1867–1934) chose for her doctoral research in December 1897. She was the wife of Pierre Curie (1859–1906), Professor of Physics at the École Municipale de Physique et de Chimie Industrielles, a Polish girl who had come from Warsaw to take her preliminary degrees in physics and mathematics at the Sorbonne, and who could now leave her three-month old daughter Irène in the care of a nurse and a grandfather while she settled down again to scientific work.[5] Her investigations were a good deal more quantitative than Becquerel's, thanks in part to the excellent electrometer which Pierre Curie had designed,[6] which she operated as a null instrument by utilizing the piezo-electric effect discovered by her husband and his brother Jacques in 1880. She used a quartz plate which could be stressed by suspending weights from it, and which under stress developed an electrostatic potential difference between opposite faces. The device could be calibrated in terms of the suspended weight; thus when Marie Curie returned the electrometer needle to zero by applying a neutralizing charge to the electrode of her ionization chambers, she knew the magnitude of the charge which had been collected on that electrode.

She began with metallic uranium, turned next to various (inactive) metals, then to pitchblende, whose radiation was more intense than that from the uranium, and so came gradually to the results which are presented systematically in the paper which follows.[7]

It is interesting to notice that substances with a high intensity of radiation are described here as "active." This is the first use of a term which has become permanent in the vocabulary of physics. The discovery of the activity of thorium had been anticipated by G. C. Schmidt of Erlangen who took some pains to point out his priority.[8] However neither he nor the Curies continued with the problem of the thorium radiation and it passed into the hands of R. B. Owens and Ernest Rutherford at McGill University in Montreal.[9] As Rutherford subse-

---

par l'uranium métallique," *ibid.*, pp. 1086–1088 (Paper 4, Vol. I); "Sur diverses propriétés des rayons uraniques," *ibid.*, *123*: 855–858; "Recherches sur les rayons uraniques," *ibid.*, 1897, *124*: 438–444; "Sur la loi de décharge dans l'air de l'uranium électrisé," *ibid.*, pp. 800–803.

[5] M. Curie, *Pierre Curie* with the addition of *Autobiographical Notes*, transl. by C. and V. Kellogg, New York: Macmillan, 1923.

[6] P.-H. Ledeboer, "Nouveaux électromètres à quadrants apériodiques," *La Lumière Électrique*, 1886, *22*: 17–20, 57–62, 145–351.

[7] I. Joliot-Curie, "Étude des 'Carnets de Laboratoire,'" in M. Curie, *Pierre Curie*, Paris: Denoël, 1955, pp. 103–112.

[8] G. C. Schmidt, "Ueber die vom Thorium und den Thorverbindungen ausgehende Strahlung," *Verhandlungen der physikalischen Gesellschaft zu Berlin*, 1898, *17*: 13, 14–16; "Ueber die von den Thorverbindungen und einigen anderen Substanzen ausgehende Strahlung," *Ann. Phys., Lpz* [3], 1898, *65*: 141–151; "Sur les radiations émises par le thorium et ses composés," *C. R. Acad. Sci.* Paris, 1898, *126*: 1264.

[9] R. B. Owens, "Thorium Radiation," *Phil. Mag.* [5], 1899, *48*: 360–387.

E. Rutherford and R. B. Owens, "Thorium and Uranium Radiation," *Transactions of the Royal Society of Canada* [2], 1899, *5*: Sec. III, 9–12.

quently demonstrated, the extraordinary activity of thorium oxide, especially in thick layers, as Marie Curie reported it here, should be ascribed to the gradual release of its emanation (Em^{220} in modern notation) into the ionization chamber.

This paper was presented for publication to the Académie des Sciences by Gabriel Lippman, Professor of Physics at the Sorbonne.—A. R.]

1

Mme. Sklodowska Curie

Rays Emitted by the Compounds of Uranium and Thorium[10]

[Translation[11] of "Rayons émis par les composés de l'uranium et du thorium," *Comptes rendus de l'Académie des Sciences, Paris*, 1898, *126*: 1101–1103 (12 April).]

I have studied the conductivity of air under the influence of the rays from uranium, discovered by Becquerel, and I have investigated whether other substances than the compounds of uranium were capable of making the air a conductor of electricity. For this study I used a parallel-plate condenser, one of the plates being covered with a uniform layer of uranium or some other substance, finely powdered. (Diameter of the plates, 8 cm; separation, 3 cm.) A difference of potential of 100 volts was established between the plates. The current across the condenser was measured in absolute value by an electrometer with a piezoelectric quartz.

I have examined a large number of metals, salts, oxides, and minerals.[12] The Table below gives for each substance the intensity of the current i in amperes (order of magnitude 10^{-11}). The substances I have studied which do not appear in the Table are at least 100 times less active than uranium.

E. Rutherford, "A Radio-active Substance emitted from Thorium Compounds," *Phil. Mag.* [5], 1900, *49*: 1–14 (Paper 5, Vol. I); "Radioactivity produced in Substances by the Action of Thorium Compounds," *ibid.*, pp. 161–192 (Paper 6, Vol. I).

[10] This work was done at the Municipal School of Industrial Physics and Chemistry.

[11] [This and all other papers originally in French or German were translated by the editor of this volume.—A. R.]

[12] The uranium used in this study was given by M. Moissan. The salts and oxides were pure preparations from the laboratory of M. Étard at the

	Amperes
Uranium lightly carburized	24×10^{-12}
Black oxide of uranium U_2O_5	27 ”
Green oxide of uranium U_3O_8	18 ”
Ammonium, potassium, sodium uranates, about . .	12 ”
Hydrated uranic acid	6 ”
Uranyl nitrate, uranous sulfate, potassium uranyl sulfate, about	7 ”
Artificial chalcolite (copper uranyl phosphate) . .	9 ”
Thorium oxide, layer 0.25 mm thick	22 ”
Thorium oxide, layer 6 mm thick	53 ”
Thorium sulfate	8 ”
Potassium fluoxytantalate	2 ”
Potassium fluoxyniobate and cerium oxide	0.3 ”
Pitchblende from Johanngeorgenstadt	83 ”
” from Cornwall	16 ”
” from Joachimsthal and from Pzibram .	67 ”
Natural chalcolite	52 ”
Autunite	27 ”
Various thorites	from 2 to 14 ”
Orangite	20 ”
Samarskite	11 ”
Fergusonite, monazite, xenotime, niobite, aeschinite	from 3 to 7 ”
Cleveite very active.	

All the compounds of uranium studied are active, and in general are more so as they contain more uranium.

The compounds of thorium are very active. Thorium oxide exceeds in activity even metallic uranium.

It is to be noticed that the two most active elements, uranium and thorium, are those which possess the greatest atomic weight.

Cerium, niobium, and tantalum appear to be slightly active.

White phosphorus is very active, but its action is probably of a different nature from that of uranium and thorium. Indeed phosphorus is not active either in the form of red phosphorus or in the form of the phosphate.

The minerals which have proved active all contain active elements. Two minerals of uranium—pitchblende (uranium oxide) and

---

School of Physics and Chemistry. M. Lacroix has kindly procured several specimens of minerals of known origin from the collection of the Muséum. Some rare, purified oxides were given by M. Demarçay. I thank these gentlemen for their assistance.

chalcolite (copper uranyl phosphate)—are much more active than uranium itself. This fact is very remarkable and leads to the belief that these minerals may contain an element which is much more active than uranium. I have reproduced some chalcolite by Debray's process using pure constituents; this artificial chalcolite was no more active than any other salt of uranum.

Absorption. The effects produced by the active substances increase with the thickness of the layer employed. This increase is very small for the compounds of uranium; it is considerable for thorium oxide, which thus appears to be partly transparent to the rays it emits.

To study the transparency of various substances, they are placed in thin sheets above the active layer. The absorption is always very strong. Nevertheless the rays penetrate metals, glass, ebonite, and paper in small thicknesses. Here is the fraction of the radiation transmitted by a sheet of aluminum 0.01 mm thick.

0.2 for uranium, ammonium uranate, uranous oxide, artificial chalcolite.
0.33 for pitchblende and natural chalcolite.
0.4 for thorium oxide and thorium sulfate, 0.5 mm layer.
0.7 for thorium oxide, 6 mm layer.

It can be seen that compounds of the same metal emit rays which are equally absorbed. The rays emitted by thorium are more penetrating than those emitted by uranium; finally, thorium oxide in a thick layer emits rays much more penetrating than those it emits in a thin layer.

Photographic impressions. I have obtained good photographic impressions with uranium, uranous oxide, pitchblende, chalcolite, thorium oxide. These substances act at small distances, whether through air, through glass, or through aluminum. Thorium sulfate gives weaker impressions and potassium fluoxytantalate impressions which are very weak.

Analogy with the secondary rays of the Röntgen rays. The properties of the rays emitted by uranium and thorium are closely analogous to those of the secondary rays of the Röntgen rays, recently studied by M. Sagnac. I have established, furthermore, that under the action of Röntgen rays, uranium, pitchblende and thorium oxide emit secondary rays which, from the point of view of the discharge of electrical bodies, have generally a larger effect than the secondary rays from lead. Among the metals studied by M. Sagnac, uranium and thorium would be placed alongside and beyond lead.

To interpret the spontaneous radiation of uranium and thorium, one might imagine that all space is constantly traversed by rays analogous to the Röntgen rays but much more penetrating, which can be absorbed only by certain elements of high atomic weight, such as uranium and thorium.

[The real interest for Marie Curie, and for Pierre Curie, who had shared in the previous work and now became a full partner, lay in the excess activity of the uranium minerals. On April 14, they ground up a 100-gram sample of pitchblende and began the hunt for an active substance whose chemical properties should differ from those of uranium and thorium, and which would probably be a new element. On July 4 there appears an entry in their laboratory notebook concerning "*Sulf, Bi, Pb, et Po*," and the discovery of polonium as it is described in the next paper was complete.[13]

This paper was presented for publication by Becquerel, who had evidently made an opportunity to become acquainted with the Curies. For the next four years he was to be the sponsor at the Académie des Sciences for all of their work. It is to be noticed that the word *radio-active* makes its first appearance in the title.—A. R.]

2

P. Curie and Mme. S. Curie

On a New Radio-active Substance Contained in Pitchblende[14]

[Translation of "Sur une substance nouvelle radio-active, contenue dans la pechblende," *Comptes rendus de l'Académie des Sciences, Paris*, 1898, *127*: 175–178 (July 18).]

Certain minerals containing uranium and thorium (pitchblende, chalcolite, uranite) are very active from the point of view of the emission of Becquerel rays. In a previous work, one of us has shown that their activity is even greater than that of uranium and thorium,

[13] I. Joliot-Curie, "Études des 'Carnets de Laboratoire,'" in M. Curie, *Pierre Curie*, Paris: Denoël, 1955, pp. 112–115.

[14] This work was done at the Municipal School of Industrial Physics and Chemistry. We are particularly grateful to M. Bémont, *chef des travaux* in chemistry, for the advice and help he has kindly given us.

and expressed the opinion that this effect was due to some other, very active substance included in small quantity in these minerals.[15]

The study of the compounds of uranium and thorium showed indeed that the property of emitting rays which make air a conductor and which act on photographic plates is a specific property of uranium and thorium, to be found in all the compounds of these metals, being smaller as the proportion of the active metal in the compound is itself less. The physical state of the substances seems to be of altogether secondary importance. Various experiments have shown that the state of mixture of the substances seems to act only as it alters the proportion of the active materials and the absorption produced by the inert substances. Certain causes (such as the presence of impurities) which act so powerfully in phosphorescence and fluorescence, are entirely without effect here. Hence it becomes very probable that if certain minerals are more active than uranium and thorium, it is because they include a substance which is more active than these metals.

We have endeavored to isolate this substance in pitchblende, and experience has confirmed the foregoing predictions.

Our chemical researches have been constantly guided by the controls afforded by the radiant activity of the products separated at each operation. Each product is placed on one of the plates of a condenser, and the conductivity acquired by the air is measured with the help of an electrometer and a piezoelectric quartz, as in the work cited above. This gives not only an indication, but a number which assesses the abundance of the product in the active substance.

The pitchblende which we analyzed was about two and a half times as active as uranium in our parallel-plate apparatus. We attacked it with acids and treated the solution obtained with hydrogen sulfide. The uranium and thorium remained in the solution. We have established the following facts:

The precipitated sulfides contained a very active substance together with lead, bismuth, copper, arsenic, and antimony.

This substance is entirely insoluble in ammonium sulfide, which distinguishes it from arsenic and antimony.

The sulfides insoluble in ammonium sulfide were dissolved in nitric acid; the active substance could be separated incompletely from the lead by sulfuric acid. By extracting the lead sulfate with dilute sulfuric acid, it is possible to dissolve in great measure the active substance entrained by the lead sulfate.

The active substance now in solution with bismuth and copper is

---

[15] Mme. Sklodowska Curie, *Comptes rendus*, vol. CXXVI, p. 1101.

completely precipitated by ammonia, which distinguishes it from copper.

Finally, the active material remains with the bismuth.

We have not yet found any definite procedure for separating the active substance from bismuth by wet methods. Nevertheless we have accomplished some incomplete separations based on the following facts:

In the solution of the sulfides by nitric acid, the portions which dissolve the most easily are the least active.

In the precipitation of the salts by water, the first portions precipitated are by far the most active.

We had observed that when pitchblende was heated, very active products were obtained by sublimation. This observation led us to a process of separation based on the difference in volatility between the active sulfide and the sulfide of the bismuth. The sulfides are heated in a vacuum in a tube of Bohemian glass at about 700°. The active sulfide is deposited in the form of a black stain in the regions of the tube which are at 250°–300°, while the bismuth sulfide remains in the hotter parts.

By these various operations, products are obtained which are more and more active. In the end we have obtained a substance whose activity is about 400 times greater than that of uranium.

We have searched among the substances known at present to see whether any of them are active. We have examined the compounds of nearly all the elements; thanks to the great kindness of several chemists we have had specimens of the rarest substances. Uranium and thorium alone are freely active; tantalum perhaps is very weakly so.

We believe then that the substance we have extracted from pitchblende contains a metal not hitherto distinguished, closely related to bismuth by its analytical properties. If the existence of this new metal is confirmed, we propose to call it *polonium* from the name of the country of origin of one of us.

M. Demarçay has kindly examined the spectrum of the substance which we are studying. He has not been able to distinguish any characteristic line outside of those due to impurities. This fact is unfavorable to the idea of the existence of a new metal. However, M. Demarçay has pointed out to us that uranium, thorium, and tantalum possess peculiar spectra, formed of innumerable very fine lines which are difficult to perceive.[16]

Perhaps we may be permitted to remark that if the existence of a

---

[16] The singularity of these three spectra is noted in the excellent publication by M. Demarçay: *Spectres électriques*, 1895.

new element is confirmed, this discovery will be due solely to the new method of investigation which the Becquerel rays provide.

[After July, there is a three-months' gap in the Curies' laboratory notebooks. The entries for November and December deal with insoluble sulfates and with the separation of an active substance from barium.[17] The standard procedure to prove the existence of a new element was to purify the working material repeatedly until it showed an unchanging atomic weight and an unchanging spectrum, both different from those of all the known elements. Since the known vacancies in the periodic table lay mostly in the neighbourhood of uranium, it was to be expected that the element the Curies were naming *radium* would be a good deal heavier than barium. Thus the atomic weight of a mixture of radium and barium ought to be greater than that of barium alone. As the next two papers show, the evidence favored the existence of radium as an element, though perhaps more credibly for its spectrum than for its atomic weight.

The final footnote to the Curies' paper indicates the immediate future of their work. From now on they would be extracting radium from larger and larger quantities of raw materials until they could accumulate the weighable quantity they needed.—A. R.]

3

P. Curie, Mme. P. Curie and G. Bémont

On a New, Strongly Radio-active Substance Contained in Pitchblende[18]

[Translation of "Sur une nouvelle substance fortement radio-active, contenue dans la pechblende," *Comptes rendus de l'Académie des Sciences, Paris*, 1898, *127*: 1215–1217 (26 December).]

Two of us have shown that by purely chemical procedures it is possible to extract from pitchblende a strongly radio-active substance. This substance is related to bismuth by its analytical properties. We

---

[17] I. Joliot-Curie, "Étude des 'Carnets de Laboratoire,'" in M. Curie, *Pierre Curie*, Paris: Denoël, 1955, pp. 115–118.

[18] This work was done at the Municipal School of Industrial Physics and Chemistry.

have expressed the opinion that perhaps the pitchblende contained a new element, for which we have proposed the name of *polonium*.[19]

The investigations which we are following at present are in agreement with the first results we obtained, but in the course of these investigations we have come upon a second, strongly radio-active substance, entirely different from the first in its chemical properties. Specifically, polonium is precipitated from acid solution by hydrogen sulfide; its salts are soluble in acids and water precipitates them from solution; polonium is completely precipitated by ammonia.

The new radio-active substance which we have just found has all the chemical appearance of nearly pure barium: it is not precipitated either by hydrogen sulfide or by ammonium sulfide, nor by ammonia; its sulfate is insoluble in water and in acids; its carbonate is insoluble in water; its chloride, very soluble in water, is insoluble in concentrated hydrochloric acid and in alcohol. Finally this substance gives the easily recognized spectrum of barium.

We believe nevertheless that this substance, although constituted in its major part by barium, contains in addition a new element which gives it its radio-activity, and which, in addition, is closely related to barium in its chemical properties.

Here are the reasons which argue for this point of view:

1. Barium and its compounds are not ordinarily radio-active; and one of us has shown that radio-activity appears to be an atomic property, persisting in all the chemical and physical states of the material.[20] From this point of view, the radio-activity of our substance, not being due to barium, must be attributed to another element.

2. The first substances which we obtained had, in the form of a hydrated chloride, a radio-activity 60 times stronger than that of metallic uranium (the radio-active intensity being evaluated by the magnitude of the conductivity of the air in our parallel-plate apparatus). When these chlorides are dissolved in water and partially precipitated by alcohol, the part precipitated is much more active than the part remaining in solution. Basing a procedure on this, one can carry out a series of fractionations, making it possible to obtain chlorides which are more and more active. We have obtained in this manner chlorides having an activity 900 times greater than that of uranium. We have been stopped by lack of material; and, considering the progress of our operations it is to be predicted that the

---

[19] P. Curie and Mme. P. Curie, *Comptes rendus*, vol. CXXVII, p. 175.
[20] Mme. P. Curie, *Comptes rendus*, vol. CXXVI, p. 1101.

activity would still have increased if we had been able to continue. These facts can be explained by the presence of a radio-active element whose chloride would be less soluble in alcohol and water than that of barium.

3. M. Demarçay has consented to examine the spectrum of our substance with a kindness which we cannot acknowledge too much. The results of his examinations are given in a special Note at the end of ours. Demarçay has found one line in the spectrum which does not seem due to any known element. This line, hardly visible with the chloride 60 times more active than uranium, has become prominent with the chloride enriched by fractionation to an activity 900 times that of uranium. The intensity of this line increases, then, at the same time as the radio-activity; that, we think, is a very serious reason for attributing it to the radio-active part of our substance.

The various reasons which we have enumerated lead us to believe that the new radio-active substance contains a new element to which we propose to give the name of *radium*.

We have measured the atomic weight of our active barium, determining the chlorine in its anhydrous chloride. We have found numbers which differ very little from those obtained in parallel measurements on inactive barium chloride; the numbers for the active barium are always a little larger, but the difference is of the order of magnitude of the experimental errors.

The new radio-active substance certainly includes a very large portion of barium; in spite of that, the radio-activity is considerable. The radio-activity of radium then must be enormous.

Uranium, thorium, polonium, radium, and their compounds make the air a conductor of electricity and act photographically on sensitive plates. In these respects, polonium and radium are considerably more active than uranium and thorium. On photographic plates one obtains good impressions with radium and polonium in a half-minute's exposure; several hours are needed to obtain the same result with uranium and thorium.

The rays emitted by the components of polonium and radium make barium platinocyanide fluorescent; their action in this regard is analogous to that of the Röntgen rays, but considerably weaker. To perform the experiment, one lays over the active substance a very thin aluminum foil on which is spread a thin layer of barium platinocyanide; in the darkness the platinocyanide appears faintly luminous above the active substance.

In this manner a source of light is obtained, which is very feeble to tell the truth, but which operates without a source of energy. Here is at least an apparent contradiction to Carnot's Principle.

Uranium and thorium give no light under these conditions, their action being probably too weak.[21]

4

Eugène Demarçay

On the Spectrum of a Radio-active Substance

[Translation of "Sur le spectre d'une substance radio-active," *Comptes rendus de l'Académie des Sciences, Paris,* 1898, *127*: 1218 (26 December).]

M. and Mme. Curie have asked me to examine with regard to its spectrum a substance containing for the most part barium chloride, in which they recognize, for reasons developed elsewhere, the presence of a new element. This substance, dissolved in distilled water weakly acidified with HCl and submitted to the action of the spark from my heavy-wire coil, furnished a brilliant spectrum which was photographed. In this way I prepared two negatives with two times of exposure, the one double the other. These two negatives, however, by the nearly equal intensity of their lines, gave identically the same result. I have measured them, and have been able to see:

1. Barium, represented with great intensity by its strong and weak lines.

2. Lead, recognized by its principal lines, which however were rather weak.

3. Platinum due to the electrodes, and the principal lines of calcium, due to the solvent.

4. A *prominent line*, stronger than the weak lines of barium, having for λ: 3814.8 (Rowland's scale). This line does not appear to me capable of attribution to any known element: first because one cannot distinguish on the negatives in question any other line than those already enumerated, except for a few weak lines of air (this excludes

---

[21] May we be permitted to thank here M. Suess, Correspondent of the Institute and Professor at the University of Vienna? Thanks to his benevolent intervention, we have obtained from the Austrian government the free gift of 100 kg of a residue from the treatment of the Joachimsthal pitchblende, containing no uranium, but containing polonium and radium. This gift will greatly facilitate our researches.

all those other elements which have at the most weak lines in the neighborhood of 3814); in the second place, in addition, because the method of purification employed for the substance eliminates precisely those elements which might cause it (Fe, Cr, Co, Ni, . . .), and which besides do not show themselves by any line, weak or strong. Moreover neither barium nor lead, as I have assured myself, gives any line whatever which coincides with it.

It was measured with reference to two lines of platinum, 3818.9 and 3801.5, which include it between them. It is near to and distinct from a bismuth line of moderate intensity.

Conclusion. The presence of the line 3814.8 confirms the existence in small quantity of a new element in the barium chloride of M. and Mme. Curie.

II. Polonium, Radiotellurium, and Confusion

[In the Curies' hands, their method looked straightforward. Radioactivity was a characteristic property of certain elements. By testing for radioactivity, one could follow the desired element through any chemical process, locate it among the precipitates and filtrates, and assess whether it was concentrating or being dispersed. The problem in applying the method lay in the fact, as yet undiscovered, that no radioactive element has a permanent existence. Each of them is continually transmuting into something different; as a result every radioactive preparation acquires new chemical properties with the passage of time. In the case of radium, whose transformation is relatively slow, the alterations were of little consequence. To concentrate radium would require nothing which the Curies had not already learned. In the case of polonium, however, as the next group of paper shows, competent chemists could arrive at flatly inconsistent conclusions, which could be resolved only by the application of Rutherford's transformation theory.

The author of the first paper in the group, Friedrich Giesel (1852–1927), was director of Buchler und Compagnie, manufacturers of quinine in Braunschweig. He was a good inorganic chemist, one of the few who succeeded in preparing his own radium and polonium, and he had loaned a radium-barium chloride preparation to his neighbors J. Elster and H. Geitel of Wolfenbüttel for some inconsequential experiments[1] to which he alludes in his opening sentence.

The changes in radioactivity which Giesel observed with radium can be explained in the following way. The immediate transformation product of radium is radon, which is occluded in the crystals of most radium salts and diffuses rapidly out of solutions. Its half-life is about four days and it transforms into an active deposit of short-lived isotopes of polonium, lead, and bismuth. Giesel was undoubtedly making his observations on the beta radiation of the radium C (or Bi^{210}) of this deposit. The total radioactivity of a freshly crystallized radium salt increases as the occluded radon and active deposit come into equilibrium with the radium. A freshly dissolved salt loses its radon and the unsupported active deposit thereupon decays rapidly.

The closing paragraphs of the paper, on polonium, are of particular interest to us, since they describe what Giesel came to regard as the

[1] J. Elster and H. Geitel, "Weitere Versuche an Becquerelstrahlen," *Ann. Phys., Lpz* [3], 1899, *68*: 83–90.

characteristics of that substance. It can be inferred that he was observing its radioactivity by the fluorescence it produced, probably in barium platinocyanide. The radiation from radium which he found would penetrate a silver thaler was perhaps the gamma radiation from the active deposit. Giesel's polonium is certainly Bi^{210}, which decays to Po^{210} with the emission of a beta ray and has a half-life of about six days. Since he could detect very little activity in a specimen two months old, it is evident that his fluorescent screen responded to beta rays and not to the alpha rays which polonium emits.

Giesel assumed, as did everyone else, that radium and polonium were as permanent as barium and bismuth; in consequence this loss of radioactivity would be puzzling. It is small wonder that he tried to regenerate it by the process which had seemed effective with radium, dissolving the preparation and recovering it from solution.—A. R.]

5

F. Giesel

Remarks on the Behavior of Radioactive Barium and on Polonium

[Translation of "Einiges über das Verhalten des radioactiven Baryts und über Polonium," *Annalen der Physik* [3], 1899, *69*: 91–94 (14 September).]

In connection with the preceding publication by Elster and Geitel, I think it advisable to add a few observations which I have made during the production of radioactive preparations from uranium ores, since they appear to me to expand our knowledge of Becquerel rays considerably.

At the same time as, and independently of, P. and S. Curie, I had isolated from the products of the manufacture of uranium salts (for which I have to thank the kindness of Messrs. de Haën of the chemical works in Hannover) a material consisting chiefly of barium sulfate, which emitted strong Becquerel rays and set a barium platinocyanide screen glowing. The material proved to be identical with the so-called radium-bearing substance prepared by Curie, although not pitchblende but other uranium ores formed the starting product. The purified chloride yielded a very effective preparation, which Elster and Geitel have used and which was exhibited to the German Physical Society.

Since it did not seem impossible that a still further enrichment of the active substance might be achieved by the employment of greater quantities than it was feasible for me to operate with, I induced the above-named factory to work up a large quantity of the ore of this material. I obtained in this way so much of the substance, which occurs in extremely small amounts in uranium ore, that I have already been able to study its properties rather more thoroughly.

I have already observed the following facts:

1. Radioactive (and thus radium-bearing) barium salts, when freshly crystallized from water, exhibit first a very small activity. This increases steadily however in the course of a few days or weeks until the effect reaches a maximum and then remains constant.

2. A concentrated solution of the active chloride in water gives at first almost the same radiation as the solid salt. After a while however the activity steadily decreases and disappears almost entirely. The crystals obtained from the solution, as already indicated under 1., gradually regain their activity.

3. All the radioactive barium salts which I have investigated (about a dozen) possess the strongest activity in the first crystallization, while successively less effective preparations are obtained from the mother liquor.

The more concentrated the liquor can be made, the more rapidly and completely the desired effect can be obtained.

4. The chloride, the bromide to a greater extent, and also the iodide, all exhibit a phosphorescence under their own rays which is independent of any previous illumination, and which is especially strong in salts dehydrated by heating the crystals. The anhydrous bromide phosphoresces very strongly with a blue-green light. In damp air it absorbs water and the phosphorescence becomes weaker; however, it can be brought back to the same intensity as often as desired by renewed heating. The phosphorescence disappears under warmth and reappears only in the cold.

This self-luminescence of the bromide appears in full strength immediately after dehydration, before the Becquerel rays have completely developed. The stronger the phosphorescence in any preparation the smaller its Becquerel radiation seems to be.

5. The green double salt of barium platinocyanide prepared from active barium chloride and potassium platinocyanide with the addition of a little potassium cyanide phosphoresces by itself very strongly, as was to be expected. The glow becomes weaker in time, however, since under the continued action of its own rays the green salt gradually goes over into the less sensitive yellow, and finally into the brown, in analogy with the behaviour of ordinary barium platinocyanide under the continued, intense action of Röntgen rays. The

green salt can be restored by solution of the brown salt and renewed crystallization.

6. A very strongly effective barium chloride, which initially was colorless, became colored with a yellowish tinge with increasing activity.

The principal question—how far the radioactivity can be increased, and whether in the end the active substance can be separated from the barium and isolated—is not yet settled. Only so much appears to be certain, that a more thorough-going improvement of the preparation than has already been obtained is not possible by fractional crystallization alone, even with the employment of far greater quantities than before.

I am still occupied with the investigation of the strongly active polonium-bearing material which is extracted at the same time as the radium but in smaller amounts. However, I have obtained a precipitate of it with hydrogen sulfide which exceeds in effectiveness the best barium preparation. Equally effective is the chloride prepared from this sulfur compound, as well as the free metal separated from the solution of the latter by metallic zinc or the galvanic current.

A remarkable difference is observed with regard to the penetrating power of the rays emitted by these two chemically different radioactive substances. While the rays from radium, for example, penetrate a silver thaler fairly well, the rays from polonium, although they are more intense, are completely held back by considerably thinner plates of metal. The silhouette of a hand, of a metal object, etc., appears on a screen more brilliantly and in higher contrast by the use of polonium rays than with the rays of radium.

Only a few experiments have been started with polonium preparations on possible changes in activity like those of the radium preparations, and I shall mention only that two small amounts of good, active hydrogen sulfide precipitates obtained two months ago have now lost their activity completely. It cannot be regenerated by renewed solution of the material and precipitation with hydrogen sulfide.

In other respects, I have been able to confirm the statements of the French investigators concerning polonium and radium.

Braunschweig, August 1899.

(Received August 5, 1899.)

[Willy Marckwald (born 1864, died in Brazil, an exile from Nazi Germany) was Professor of Chemistry at the University of Berlin, and an experienced and resourceful inorganic chemist. As he tells in the following paper, he was drawn into the study of radioactivity as a consultant for the manufacturing firm of Dr. Richard Sthamer in Hamburg, and decided to work on the radioactive material which was found with bismuth in uranium minerals. The electroscope which he used to test for radioactivity would be far more sensitive to alpha than to beta radiation, so he missed the Bi^{210} which Giesel had worked with and followed Po^{210} instead. Thus he had little difficulty in showing that his radioactive material was chemically different from bismuth. Since Po^{210} has a half-life of about 140 days, he could reasonably claim that he had noticed no diminution in the radioactivity of his material over a period of "several months" by the qualitative tests he used, although in fact the loss in radioactivity must have amounted to about 30%.

Late in 1899 the Curies had discovered that a plate of any material held near a preparation of radium chloride would acquire a temporary radioactivity which died away over the course of a few hours. The effect is actually produced by the diffusion of gaseous radon away from the radium chloride and its subsequent transformation to Po^{214}, which is not gaseous at ordinary temperatures and so deposits on the first solid it touches. The Curies convinced themselves however that what they obtained was not a material deposit, and consequently ascribed the radioactivity to a process of induction by the primary radioactivity of the radium.[2] By an extension of their conclusion, the idea was rather generally accepted that any temporary radioactivity resulted from an induction of this kind. Under such an interpretation, neither the Curies' polonium nor Giesel's would be an element since neither possessed a "permanent" radioactivity.—A. R.]

6

W. Marckwald

On Radioactive Bismuth (Polonium)

[Translation of "Ueber das radioactive Wismuth (Polonium) [Vorläufige Mittheilung]," *Berichte der deutschen chemischen Gesellschaft*, 1902, *35*: 2285–2288.]

(From the IInd Chemical Institute of the University. Presented at the session [of June 9, 1902] by the author.)

Inspired by Becquerel's beautiful observations on the characteristic radiation of uranium minerals and uranium preparations, P. and S.

---

[2] P. Curie and Mme. M.-P. Curie, "Sur la radioactivité provoquée par les rayons de Becquerel," *C. R. Acad. Sci., Paris*, 1899, *129*: 714–716.

Curie[3] in 1898 undertook a thorough chemical investigation of the pitchblendes. In this way they found, in extremely small quantity, two exceedingly interesting constituents of these minerals, which proved to be radioactive to a much higher degree than the compounds of uranium.

The first of these newly discovered substances showed all the chemical reactions of bismuth, and was distinguished from it only by its radioactivity, which was about a hundred times greater than that of the pitchblende. By partial sublimation of the sulfide, partial precipitation of the nitrate by water, etc., the metal in the form of the compound in question could be separated into derivatives of inactive bismuth on the one hand, and on the other into compounds of radioactive bismuth which were approximately four times stronger. Although the discoverers had no doubt that the end product still consisted in very great part of bismuth, they were not able to bring about any further enrichment of the active constituent. In the meantime, they have proposed the name polonium for the unknown element which they suppose to be the cause of the radioactivity.

Immediately after the discovery of radioactive bismuth followed that of radium. The authors have been more fortunate, as is well known, in their attempts to separate this substance from the barium which accompanied it. By crystallization of the chlorides, they have succeeded in freeing radium almost entirely from barium, if not in preparing it in absolute purity. How the barium can be removed to any desired degree from radium-bearing barium chloride by a suitable process of crystallization, I have recently described.[4]

Further investigation of these radioactive substances, to which still others have been added—actinium, europium, radioactive lead—has allowed a remarkable difference to appear between the activity of radium and polonium. Although the activity of freshly refined radium preparations grows rapidly at first and finally becomes constant, F. Giesel[5] observed that his radioactive bismuth lost the greater part of its strength after only a few weeks, and the Curies' preparation shows the same phenomenon, although slower. This led Giesel to conjecture that "We have in polonium perhaps nothing more than an induced activity of bismuth." It has repeatedly been established that a radioactive substance can transfer its activity to another substance, especially if the latter is precipitated from a solution containing an active material.

Recently the discoverers of polonium seem also to have acceded to

[3] *Compt. rend. 127,* 175 (1898).

[4] *Chem. News 84,* 190 (1901).

[5] Cf. in particular: F. Giesel, Ueber radioactive Substanzen und deren Strahlen. Ahrens' *Sammlung chem. und chem.-techn. Vortr. 7,* 1 (1902).

Giesel's point of view. At least, a recently published paper by P. and S. Curie[6] contains this sentence in a footnote: "Polonium is a species of active bismuth; it has not yet been proved that it contains a new element."

Observations which I have made during an investigation of radioactive bismuth have already somewhat clarified the nature of this substance. To secure for myself the uninterrupted prosecution of these investigations in the manner I have chosen, I shall sketch the preliminary results here briefly.

In the spring of last year, several kilos of an intermediate product in the working-up of the pitchblende ore of uranium were handed over to me for investigation by the firm of Dr. Richard Sthamer in Hamburg. According to the information given me, the material consisted of the residue left after the decomposition of the Joachimsthal pitchblende by sulfuric acid and its leaching by water. The goal of the investigation was first to establish whether any radioactive substance might be obtained from the product, and then to work out a technically feasible process for the extraction of this substance.

The material proved to be relatively rich in radioactive bismuth. By established procedures, bismuth oxychloride was extracted amounting to about one percent by weight of the raw material. The product was very strongly radioactive. *No diminution of its activity has been noticed over the course of several months.*

After a good many attempts to isolate a radioactive constituent from this material, which were unsuccessful in part, and in part gave only slightly satisfactory results (and these have been abandoned), I have finally succeeded in solving this problem in the simplest possible manner.

On the assumption that radioactive bismuth contains in addition to ordinary bismuth a second radioactive metal which stands very close to the former in its chemical reactions, the behavior of the salts in electrolysis was investigated. If the two metals exhibit any potential difference—and that is to be expected in close analogy with ordinary chemical behavior—this should be observable in electrolysis, in the sense that in the first metal laid down, either the active or the inactive constituent will be enriched. At the first attempt, under quite arbitrarily chosen conditions, it turned out *that the metal first laid down showed a very much stronger activity than the original material.* This result suggested an attempt to displace the ions of the radioactive metal in a hydrochloric acid solution of the chloride directly by bismuth ions, by dipping a polished bismuth rod into the solution.

---

[6] *Compt. rend. 134,* 85 (1902). [This is reprinted as Paper 10 in Vol. I.—A. R.]

According to expectation, the metal was covered immediately with a fine black coating, which visibly increased in a few hours. When it was taken from the solution and washed with hydrochloric acid and alcohol, the rod had an astonishing effect upon an electroscope. At the distance of a decimeter, the leaves of the charged electroscope clapped together in an instant, indeed a stout gutta-percha rod which had been vigorously rubbed with a fox's tail was completely discharged on its mere approach.

It is of the greatest significance that the whole of the radioactive metal is deposited on the bismuth rod in the course of a few days. When 8 gm of bismuth oxychloride in hydrochloric acid solution were allowed to stand three days with a plate of bismuth, the salt in the solution was as good as inactive. For a more thorough test, a fresh bismuth rod was allowed to stand in this solution for 24 hours. This rod remained completely clear and showed only a very weak radioactivity.

The deposit could be scraped off the bismuth plate very easily, and apparently completely. It weighed about 5 mg. Hence, at the highest, the radioactive bismuth contained 1/10 per cent of the radioactive metal. A ton of pitchblende ore then, at a rough estimate, contains no more than 1 gm of it.

The deposit is not pure metal, but probably contains some chlorine. On heating, a small part of it evaporated, very much after the habit of a chloride. The remainder fused together into a white, shiny grain of metal, which is extremely brittle. It dissolves easily in nitric acid. The solution shows, as much as it has been tested so far, the characteristic reactions of bismuth. However, the metal must differ from this, since the displacement of a metal from solution by itself—except by the formation of concentration currents which are excluded in this case—would contradict the principle of the conservation of energy.

The rays emitted by the metal—and quite as strongly by the salts also—differ characteristically from the rays of radium. In particular, they seem incapable of penetrating any obstacle. A bismuth rod prepared by the foregoing process, one which shows the strongest effect on an electroscope, need only be wrapped with a little filter paper to eliminate its activity almost entirely.

The small penetrating power of the rays is also the reason why one cannot obtain much stronger effects by the use of deposits on the bismuth thicker than a few ten-thousandths of a gram. The firm of Dr. Richard Sthamer intends to bring such rods onto the market for the purpose of physical research and demonstration.

My next task will be to obtain enough of the new metal so that a determination of atomic weight can be carried out. The firm to which I have referred, which has promoted my investigation in the friendliest manner, has extracted bismuth at its plant—and radium as well—from a great quantity of pitchblende residues, according to my specifications. To my deepest gratitude, it has obligingly turned over to me almost a kilo of bismuth oxychloride. I hope with this amount to be able to solve the problem I have set myself.

[Marckwald was enthusiastic about his new radioactive material and seized every occasion to talk about it. He had presented the previous paper in person to the Deutsche chemische Gesellschaft on June 9, 1902. He read another with essentially the same content to the Deutsche physikalische Gesellschaft on June 27,[7] and on September 23 he delivered a demonstration lecture to the 74th Versammlung deutscher Naturforscher und Ärzte meeting that summer at Karlsbad.[8] Although he remained scrupulous in giving credit to the Curies for their prior work, he clearly implied that his radioactive substance was different from theirs and more likely to prove a genuine element.

When the Curies had voiced their doubts that polonium might be an element, it had been in an entirely different context. At that moment they had been concerned with what they thought to be rash speculation on Becquerel's part—that the emission of radiation might produce chemical changes in a radioactive substance.[9] Now that polonium seemed about to be pre-empted by Marckwald, Marie Curie came to its defense in the paper which follows, published to meet Marckwald's audience in the same journal as his last, the *Physikalische Zeitschrift*. It contains nothing new, and it is perhaps unfortunate that she did not try with her own polonium Marckwald's most significant innovation, the displacement of the material from solution by bismuth. It would have shortened the controversy which followed.—A. R.]

[7] W. Marckwald, "Ueber das radioactive Wismuth (Polonium)," *Verh. dtsch phys. Ges.*, 1902, *4*: 252–254.

[8] W. Marckwald, "Das radioactive Wismuth (Polonium)," *Phys. Z.*, 1902–03, *4*: 51–54.

[9] H. Becquerel, "Sur la radioactivité de l'uranium," *C. R. Acad. Sci., Paris*, 1901, *133*: 977–980. [This is reprinted as Paper 9 in Vol. I—A.R.]

7

Mme. Curie

On the Radioactive Substance "Polonium"

[Translation of "Über den radioaktiven Stoff „Polonium“," *Physikalische Zeitschrift*, 1902–03, *4*: 234–235.]

In the issue of this journal for October 10, 1902, there appeared a paper by Marckwald, "Radioactive Bismuth (Polonium)." This paper prompts me to the following remarks.

Marckwald ascribes to me and to M. Curie the notion that our polonium contains no new element. Marckwald draws this conclusion from the following sentence which he finds in one of our papers: "Polonium is a species of active bismuth; it has not yet been proved that it contains a new element."[10]

Marckwald gives these words a meaning which they in no way possess. We only wished to say that polonium possesses the chemical properties of bismuth and that we had not yet succeeded in demonstrating its difference from that metal.

In the course of our researches on radioactive substances we have become convinced that radioactivity offers a new means of investigating unknown elements. Following this idea, we have discovered two radioactive substances of extraordinary strength, and have supposed the existence in them of two new elements, polonium and radium.[11] This opinion of ours was at first hardly widespread; few investigators believed in the existence of the new radioactive elements. It was our task to establish the correctness of our starting point. The proof arrived for radium. It was supplied by the spectrum analysis, by the purification of the chloride, and by the determination of the atomic weight of the new element.[12] On the basis of this work it seems permissible to assume that any radioactive substance which is chemically different from the known radioactive substances contains a new element. The phenomenon of induced radioactivity which we have discovered[13] shows, however, that this rule cannot be applied without care. This is especially so for substances whose radioactivity

[10] Curie, *Compt. rend.*, *134*, 85 (1902). [This is reprinted as Paper 10 in Vol. I—A. R.]

[11] *Compt. rend.*, July 1898 and December 1898.

[12] Demarçay, *Compt. rend.*, December 1898, November 1899, July 1900; Curie, November 1899, August 1900, July 1902.

[13] Curie, *Compt. rend.*, 6 November 1899.

diminishes with time, as is the case with our preparations of polonium. And since we have not yet been able to show that polonium is a new element, we are not yet entitled to assert it.

I have worked a good deal on polonium, and have obtained pure, highly active preparations (for example 30,000 times more active than metallic uranium). These preparations have been studied by Demarçay. He has noticed no new lines in the spectrum beyond the lines of bismuth. Nevertheless Demarçay believes that the material does not consist exclusively of bismuth.

I have steadily shared this opinion. Although polonium comes very near to bismuth, it is still possible to concentrate the radioactive substance. To this end, as is known, there serve: 1. fractional precipitation of a nitric acid solution with water, the precipitate being more active than the dissolved material; 2. fractional precipitation of a strongly acidic solution with hydrogen sulfide, the sulfides first deposited being the most active; 3. sublimation of the sulfide in a vacuum, the active sulfide being more volatile than the inactive.[14] As the concentration becomes greater the substance seems to show a steadily greater difference from bismuth. If a nitric acid solution is precipitated with water, then nearly every time a precipitate forms which dissolves neither in dilute nor in concentrated acids; these precipitates are white or tinged with yellow to red-brown. I have never obtained such precipitates either with pure bismuth or with a bismuth on which activity has been induced; they form only with concentrated preparations of polonium.

It is possible that polonium does not possess a sensitive spectrum reaction. It would be necessary then to determine the atomic weight of the metal in a pure and very concentrated preparation. I have not done this yet since the production of a concentrated preparation is difficult. I should be happy if Marckwald's methods might lead easily to that goal.

Nevertheless I have not abandoned the hope of producing pure polonium, and shall continue to work at it when I have a sufficient quantity of radioactive bismuth at my disposal.

Marckwald's polonium appears to be the same as ours. It accompanies the bismuth which is derived from uranium ores and emits rays that are absorbed exceptionally easily, as our preparations do. The difference lies in the fact that Marckwald's polonium undergoes no decrease in activity. However, if the decrease in activity proceeds as slowly as with our preparations, then over the course of a few months it can be established only by measurement. For example one of our polonium preparations lost half of its activity in eleven

---

[14] Curie, Congrès de Physique, 1900.

months and after this lapse of time still showed a strong discharging and photographic effect. The decrease in activity must be measured on pure preparations. Radioactive bismuth derived from uranium ores can contain traces of radium. I do not know whether Marckwald's measurements of activity were carried out on pure preparations over several months.

We possess metallic, polonium-bearing bismuth which remains very active.

It is very much to be wished that works on radioactive materials might contain numbers to give the activity of the substances referred to. To this end one can use the electrical conductivity which is excited in the air by these substances in definite proportions. The action of a substance can be compared with that of metallic uranium. It would be still better to use for comparative measurements radium-bearing barium salts of a known concentration (i.e. those which yield a definite atomic weight). This will become possible with a wider distribution of radium preparations.

Such estimates of activity are arbitrary, to be sure, but they offer a valuable means of comparing the results from different pieces of work.

(Received 12 December 1902)

[Marckwald's next advance, described in a paper which is omitted here,[15] was to extract his radioactive material from solution by a reaction characteristic of tellurium. This prompted him to name his new substance *radiotellurium*, by which he intended to imply first that the material was not tellurium although it resembled it, and second that it was certainly not polonium. In the paper which follows, he reiterates this point with a good deal of polemic vigor.

From the beginning, Marckwald had been strongly impressed by the radiation from his new substance, which was extraordinarily intense as gauged by its ionization, its photographic effects, and its power to produce phosphorescence, and yet had so little penetrating power that it could be stopped by a sheet of tissue paper or even the lacquer coating on a barium platinocyanide screen. At the Versammlung deutscher Naturforscher und Ärzte the previous summer, it had been brought to his attention that what Giesel called polonium differed from the polonium of the Curies in the rays it emitted.[8] Marckwald's natural concern was with chemical properties, but it is interesting to notice in the next paper, in his discussion on polonium, that he does draw conclusions from the type of radiation

---

[15] W. Marckwald, "Ueber den radioactiven Bestandtheil des Wismuths aus Joachimsthaler Pechblende," *Ber. dtsch chem. Ges.*, 1902, *35*: 4239–4240.

emitted, and that he has adopted (in 1903) Rutherford's nomenclature of alpha and beta rays.

The modern reader will find a clue to the interpretation of the various conflicting claims in this distinction between the characteristic rays. Marckwald kept his attention on the alpha rays, so his radioactive substance must be Po^{210}. Giesel, on the other hand, was watching for beta rays and so would be working with Bi^{210}. Thus both Marckwald and Giesel were correct when they extracted "radiotellurium" from "polonium," although without the transformation theory they would necessarily draw conflicting conclusions from the facts.

Rather more confusing is Marckwald's extraction of a temporarily active tellurium from radium chloride. In the solid state, the radium chloride would occlude its emanation, and so contain also the first transformation product of the emanation Po^{218}, which emits alpha particles. This isotope has a half-life of three minutes, which means that it would show in the earliest stages of the extraction and disappear before the process was completed, exactly as Marckwald reports.

A calculation from their half-lives shows that a uranium ore contains about 2400 times as much radium as polonium. There was little real chance then that Marckwald would ever obtain enough of his radiotellurium to bring it to a decent degree of chemical purity.—A. R.]

8

W. Marckwald

On the Radioactive Constituent of the Bismuth from the Joachimsthal Pitchblende. III[16]

[Translation of "Ueber den radioactiven Bestandtheil des Wismuths aus Joachimsthaler Pechblende. III," *Berichte der deutschen chemischen Gesellschaft*, 1903, *36*: 2662–2667.]

(From the IInd Chemical Institute of the University of Berlin, presented by the author at the session of July 13 [1903].)

I. *On Radiotellurium*

In earlier communications, I have shown that with a hydrochloric acid solution of the radioactive bismuth chloride which is obtained

[16] Cf. this Journal *35*, 2285 and 4239 [1902]: see also *Chemiker-Ztg*. *26*, 895 [1902].

from the Joachimsthal pitchblende, the radioactive component can be separated from the bismuth by two different methods. When metallic bismuth or antimony is dipped in the solution, the active constituent, mixed with a good many other substances, is deposited on the metal; it is precipitated in a purer state by stannous chloride as a fine, black deposit. Because of the complete agreement of the latter product with tellurium in all of its properties—radioactivity excepted—I have provisionally designated it as radiotellurium.

By processing 6 kilos of bismuth oxychloride which were derived from about 2000 kilos of pitchblende, I have obtained 1.5 g of this radiotellurium. Further investigation has shown that it consists almost entirely of *ordinary tellurium*, so that the radioactive constituent amounts to only a fraction of a per cent. For the separation of the tellurium the metal was converted to the chloride, and the tellurium precipitated from not too strong a hydrochloric acid solution by hydrazine hydrochloride.[17] This is extremely effective; a single repetition of the precipitation is enough to produce almost inactive tellurium.

The filtrate contained the active substance, still contaminated with bismuth, tin, and some tellurium. When the solution was concentrated and treated with a few drops of stannous chloride, a very scanty, dark precipitate appeared after digestion on a waterbath, and could be collected on a filter. Its weight amounted to 4 milligrams. I am by no means convinced that this product is completely homogeneous. Considering the great cost of the material, I have thought it better to refrain from any further chemical investigation until there is the possibility of obtaining a greater quantity.

The filtrate from the precipitation by stannous chloride still contains, even when the filtration has been repeated several times, small quantities of the active substance, perhaps in a colloidal solution. The following observations speak for this. If to a part of the filtrate, a few tenths of a milligram of telluric acid are added in an aqueous solution, the tellurium precipitation which is produced by the excess stannous chloride entrains the active substance completely. If a stick of bismuth is dipped into another part of the filtrate, it becomes only weakly active, even after long-continued action; if however a few drops of bromine water are added first, to oxidize the stannous chloride and convert some of the colloidal metal into the bromide, the active substance deposits completely on the stick of metal, which becomes highly active. This effect is very much worth considering if the active substance is to be extracted from a solution which contains a good deal of bismuth but no tellurium, as will be discussed below.

---

[17] A. Gutbier, this journal *34*, 2724 [1901].

The active substance dissolves easily in cold, dilute nitric acid. When the acid is evaporated and the residue taken up with hydrochloric acid, a solution of the chloride is obtained from which the active substance can be deposited in a state of fine division on metal plates of copper, tin, antimony, etc. dipped into it. To characterize the effectiveness of these deposits, I shall add only that on a copper plate of 4 sq cm surface a deposit of about 1/100 mg is sufficient, when it is brought close to a zinc blende screen, to make its glow visible in an auditorium with more than a hundred people.

Even if this substance is contained in pitchblende in much smaller quantities than radium, its extraction will involve smaller difficulties once a suitable way has been discovered, since chemical reactions are available which at the present time are still lacking for the separation of radium from barium.

To obtain the expensive starting material which served for the carrying out of this investigation, the Royal Academy of Sciences has made me a grant, for which I express my thanks.

II. *On Polonium*

My last paper was followed by a number of publications with the above title, whose contents I must go into briefly. Madame S. Curie[18] protests that her remark, "Polonium is only a species of active bismuth," was interpreted as though she had given up seeking for a new element in polonium; she describes some of the reactions she has observed recently which show polonium, enriched by her process of fractionation, diverging from bismuth. As concerns the first point, it is superfluous for me to emphasize that it lay far from my thoughts to diminish the immortal merit which the Curies, husband and wife, have earned for the discovery of the new radioactive elements when I cited that sentence.

Madame Curie's new observations on the properties of her enriched polonium confirm the conjecture which I had previously entertained that this substance is in no way identical with radiotellurium. On precipitation of the nitric acid solution by water, polonium yields deposits which are white or yellow-to-brown in color and insoluble in acids. This differentiates it thoroughly from radiotellurium, and constitutes a more important difference than the inconstancy of the radiating power of polonium. Whether the Curies' polonium does not perhaps contain radiotellurium is a question which must be left to the discoverers of polonium to test.

Two papers on this subject by F. Giesel have recently appeared. The first[19] confirms in essence my previous observations, insofar as

[18] *Physikalische Zeitschrift 4*, 234.
[19] This journal *36*, 728 [1903].

they relate to the production of a radioactive deposit on sticks of bismuth.

In a footnote the author alludes to the way in which platinum, gold, and palladium become radioactive in a slight degree when dipped in a polonium solution, but leaves the question still open whether reactions analogous to those of bismuth are exhibited. This effect, which I had previously observed for platinum, and which tellurium shows in addition, arises from an entirely different cause; this is shown by the fact that none of the substances named, even on long contact with the solution, diminishes perceptibly the activity of the dissolved material, while bismuth eliminates the carrier of the activity completely from the solution.

Giesel was unable to detect tellurium in the polonium, or to obtain the stannous chloride reaction. This failure arises only from a lack of tellurium. I have purchased "polonium" from the quinine works in Braunschweig, which Hr. Giesel directs. A gram of its oxychloride was dissolved in hydrochloric acid and treated with a few drops of stannous chloride solution. Even after long digestion no visible precipitate appeared, but when the solution was filtered a distinct brown coloring could be seen on the filter. After washing, the filter was radioactive in the highest degree, and fully equal to my strongest preparations. The filter had a strong odor of ozone, and disintegrated completely in the course of a few weeks. To the filtrate, as described above, a trace of telluric acid was added to bring down the rest of the radioactive material. When the bismuth was now removed from the solution it showed only a little α-radiation; the β-radiation, however, seemed to undergo no important decrease.

In a very recent paper, *On Polonium and the Inducing Power of Radium*,[20] Giesel brings forward an observation which seems to him capable of placing the polonium question in a new light. When he dipped a stick of bismuth into a radium solution, he obtained a strongly α-radiating metal. Now, if Giesel is to relate this observation to mine, the implication is that the activity in the two cases may have the same cause. This is in no way the case.

The experiment described by Giesel is one which I made a long time ago. Certainly I do not possess anything like as strong a radium preparation as Giesel. But even by the use of a radium preparation of about 1 per cent radium content a piece of bismuth dipped in the solution acquires a strong radioactivity. The strength of the radiation, to be sure, is not comparable with that of my strongest radiotellurium. Obviously this phenomenon is completely analogous

[20] This journal *36*, 2368 [1903].

with that discussed above, which Giesel and I have observed when noble metals were dipped in polonium solutions. In both cases the active constituent was not perceptibly removed, certainly not totally removed from the solution.

Nevertheless, Giesel's latest communication has induced me to start a new experiment which was easily feasible and well-adapted to confirm my understanding of the situation. 0.01 g of radium chloride (of about 2.5 per cent) were treated with 0.2 mg of telluric acid in a weak hydrochloric acid solution and the tellurium precipitated with stannous chloride. As was to be expected, a very active tellurium appeared. However all precipitates from a radium-bearing solution are more or less active, in spite of the best washing. But when the metal was dissolved from the filter with nitric acid, converted to the chloride, and deposited on copper from very dilute solution, the activity was only just detectable by the electroscope. In this respect the induced (?) activity of the tellurium is completely different from that of radiotellurium.

In conclusion, I must go briefly into one sentence from Giesel's paper, in order that its contents should not be considered as accepted by me. Giesel writes:

> Marckwald considers the deposit formed on the bismuth, in part at least, as metallic polonium, and gives as his opinion that the electrolytic separation is a proof of the presence of an element different from bismuth (more electronegative) and standing close to tellurium.

Already in a previous paper[21] I have emphasized to Hr. Giesel that I do not claim the name of polonium for the radioactive substance which I have extracted. This name, which was introduced at the time of the discovery of radioactive bismuth, involves a conception so indefinite that it was necessary to distinguish between the Curie polonium and the Giesel polonium. It would be out of place to designate still a third substance as polonium. For that reason I have provisionally called the substance which I have extracted "radiotellurium." In that regard, I have never even proposed that I consider its close relationship with tellurium as *proved.* I have emphasized at every opportunity that this relationship was merely a *conjecture* of mine. This conjecture has rendered me the best service, since it has shown me the way to extract the active substance, to separate it not only from bismuth but also from tellurium. Nevertheless I am a long way from wishing to draw from these facts any binding conclusions as to the nature of a substance which I do not believe I have once held in my hands in a pure state.

---

[21] This journal *35*, 4240 [1902].

III. Radium F, Radiotellurium, and Polonium

[The controversy over polonium is less interesting in itself than in the way it involved such gifted investigators as Marie Curie, Giesel, and Marckwald. There were two difficulties which plagued every radiochemical investigation, and which were more perplexing as they were less expected. The radioactive substance was usually present in extremely minute proportions, and by itself would often fail to precipitate. It was necessary to entrain it with another element, which might be present as an impurity in the early stages of an extraction but in later stages had to be added deliberately, as Marckwald had added tellurium to radium. In addition, every radioactive substance was undergoing steady transmutation, so that the chemical properties associated with the radioactivity were apt to change with the age of the preparation.

Clear recognition that transmutations could occur first came in 1902, at about the time that Marckwald took up the study of radioactivity. In July of that year there was published in England a paper by Ernest Rutherford and Frederick Soddy of Montreal.[1] The evidence it advanced seemed slender enough—a few years of work by Rutherford on some temporary radioactivities connected with thorium and a few months of work by Soddy on the behaviour of freshly precipitated thorium hydroxide. Nevertheless the two young men had convinced themselves, and the results of their continuing experiments only confirmed the hypothesis that radioactivity was the external sign of atoms in transformation.[2] By the end of 1903, Rutherford knew that radium transformed itself into a gas which he called *emanation*, and that emanation in transforming laid down a solid deposit within which at least three other transformations took place.[3]

---

[1] E. Rutherford and F. Soddy, "The Radioactivity of Thorium Compounds. II. The Cause and Nature of Radioactivity," *J. chem. Soc.*, 1902, *81*: 837–860 (Paper 11, Vol. I).

[2] E. Rutherford and F. Soddy, "The Cause and Nature of Radioactivity. Part I," *Phil. Mag.* [6], 1902, *4*: 370–376; "The Cause and Nature of Radioactivity. Part II," *ibid.*, pp. 569–585 (a selection is reprinted as Paper 12, Vol. I); "The Radioactivity of Uranium," *ibid.*, 1903., *5*: 441–445; "A Comparative Study of the Radioactivity of Uranium and Thorium," *ibid.*, pp. 445–457; "Condensation of the Radioactive Emanations," *ibid.*, pp. 561–576; "Radioactive Change," *ibid.*, pp. 576–591 (Paper 13, Vol. I).

[3] E. Rutherford and H. T. Barnes, "Heating Effect of the Radium Emanation," *The Physical Review*, 1904, *18*: 118–120; *Phil. Mag.* [6], 1904, *7*: 202–219; *Nature, Lond.*, 1903, *68*: 622.

The existence of this interpretation of radioactivity was brought sharply to Marckwald's attention in February, 1904. Soddy, who was spending the winter in Ramsay's laboratory in London, took occasion to comment on Marckwald's work in a letter to the English scientific weekly *Nature*.[4] Soddy was inclined to identify radiotellurium with polonium, since they shared the common property of emitting only alpha rays. In addition, he pointed out that Marckwald's claims were inconsistent. If radiotellurium was highly active, it could not also be permanent. Each ray emitted marked an atom transmuted, and thus the stronger the rays, the more rapid must be the transmutation. Marckwald was confident of his chemistry, but he saw the force of Soddy's remarks, and he took steps to measure the activity of his radiotellurium to see whether it was permanent or actually died away in time.

This was to be the decisive action. Rutherford had shown that the time required for a substance to lose half its radioactivity—the *half-life*, in modern terms—was an unalterable and completely characteristic property of the substance. Now, through the rest of 1904, while the two young physicists whom Marckwald had recruited made their measurements, Rutherford began to connect radiotellurium with radium. He found evidence for two more substances in the transformation chain which began with radium emanation. Since he was now designating the successive members of the chain by the letters of the alphabet, they became radium D and radium E. Radium E emitted alpha rays, and he found that they possessed the same power of penetration as those from a specimen of radiotellurium he had purchased from Marckwald's sponsor, Sthamer of Hamburg. From the relative intensity of its radiation, he estimated the half-life of radium E at about one year.[5]

As can be seen in the paper which follows, Marckwald adopted the transformation theory whole-heartedly. Indeed it was to his interest, since it guaranteed, through the exponential law of decay, that his radiotellurium contained only a single active substance, however much tellurium might still be mixed with it. It is also to be noticed that Marckwald's chemistry would put radiotellurium in its proper place in the periodic table, in the column just to the left of the halogens, below selenium and tellurium.—A. R.]

---

[4] F. Soddy, "Radio-tellurium," *Nature, Lond.*, 1903–04, *69*: 347, 461–462. W. Marckwald, "Radio-tellurium," *ibid.*, p. 461.

[5] E. Rutherford, "Slow Transformation Products of Radium," *Phil. Mag.* [6], 1904, *8*: 636–650; "The Succession of Changes in Radioactive Bodies," *Philosophical Transactions of the Royal Society of London, A*, 1905, *204*: 169–219 (Paper 16, Vol. I).

9

W. Marckwald

On Radiotellurium. IV[6]

[Translation of "Ueber das Radiotellur. IV," *Berichte der deutschen chemischen Gesellschaft*, 1905, *38*: 591–594.]

(From the IInd Chemical Institute of the University of Berlin. Received January 24, 1905; presented at the session [of January 23, 1905] by the author).

Through the great generosity of the proprietor of the chemical works of Dr. Rich. Sthamer in Hamburg a quantity of raw tellurium was placed at my disposal; it had been extracted from 5 tons of the residues from the Joachimsthal uranium installation, corresponding to about 15 tons of pitchblende. The extraction was undertaken essentially as I have specified before, that is to say by precipitation with stannous chloride from the solution of a previously prepared bismuth chloride.

This precipitate still contains a great many impurities. For purification it was dissolved in dilute nitric acid, and after filtration the solution was evaporated; to remove the nitric acid, the residue was repeatedly evaporated with hydrochloric acid; it was then taken up with dilute hydrochloric acid, and sulfurous acid was added to that solution. At this point a precipitate formed which consisted of a mixture of selenium, tellurium, and radiotellurium. Hence it follows that the chloride of radiotellurium is also reduced by sulfurous acid, while, as has been previously shown, it is distinguished from tellurium chloride by its behaviour toward hydrazine.

In the precipitation by sulfurous acid it was observed that radiotellurium reached the point of separation with the greatest relative difficulty. Since from a solution containing selenium and tellurium, the selenium is always precipitated first by the reducing agent, radiotellurium would fit well into the succession of the periodic system.

The weight of the whole precipitate amounted to 16 g. To extract the radiotellurium from the selenium and tellurium, it was more profitable to use another method than the separation by hydrazine previously indicated. The oxide of radiotellurium, as might be anticipated on the assumption that it occupies its supposed place

[6] This journal *35*, 2285 and 4239 (1902); *36*, 2662 (1903).

in the periodic system of the elements, does not have the properties of an acid anhydride. As is known, telluric acid is so weak an acid that it forms no stable ammonium salt, but dissolves easily and copiously in excess ammonium hydroxide. In contrast, the oxide of radiotellurium is completely insoluble in ammonia.

To extract the radiotellurium, the precipitate obtained with sulfurous acid was dissolved in dilute nitric acid, the solution evaporated to dryness and the residue heated with ammonia. From it there remained only a very scanty residue which was collected on a filter. Its weight came to about 3 mg. However it exhibited the total yield of radiotellurium, for the tellurium extracted from the solution was by comparison only very weakly radioactive.

It is self-evident that the strength of the substance collected on the filter was enormously great. It exceeded in purity many times over—to judge from the yield—the product previously described. However even this substance offers no guarantee of complete purity, although on account of the cost of the material any further chemical investigation must be renounced.

In spite of the insignificant quantity in which radiotellurium is contained in pitchblende, its extraction by the method here described is so simple and sure that its production is not altogether too difficult. Since for most physical research a thousandth of a milligram would be enough, and for all purposes of demonstration a hundredth of a milligram, the demand for this substance, setting aside the problem of its chemical investigation, ought not to be difficult to meet.

For almost a year I have been occupied in examining the question whether radiotellurium maintains its activity continually, or whether its action decreases. I thought I had noticed the latter, earlier, on the sticks of bismuth coated with the material. However since these sticks, according to the present state of our knowledge, held only thousandths of a milligram of active material, the decrease in their action might be ascribed to mechanical influences.

The precipitates obtained later with the pure substance were so strong that no weakening could be noticed on rough tests. The copper plate described in my last communication on this subject, which carries hardly 1/100 mg of radiotellurium, after two years of use still excites a phosphorescent screen sufficiently to make its glow easily visible at a distance of 10–15 meters.

Exact investigations however have exhibited with certainty the decrease in the action of radiotellurium. Thereby the disintegration theory of Rutherford and Soddy, according to which a radioactive element must lose activity more rapidly the stronger it is, has undergone a new confirmation.

The results of this investigation in which Dr. Greinacher and Hr.

Herrmann took part, and which was carried out in the Physical Institute of this University, will be thoroughly reported in another place. Considering however that Dr. Stefan Meyer and Dr. Egon von Schweidler have communicated an investigation treating the same question to the Imperial Academy of Sciences in Vienna at the session of December 1 of the year past, as I have learned from the reprint which the authors kindly sent me from No. XXV of the *Anzeiger* of the Academy, I shall here make known provisionally the result of our experiments.

It should be remarked beforehand that our experiment was carried out with the most extremely small quantity of radiotellurium (of the magnitude at the most of 1/1000 mg), which was purified by the process described above and deposited on a little plate of silver, whereas Meyer and von Schweidler used commercial sticks of bismuth, or copper plates. Their experiments extended over 3 months, ours over 10 months. There is nothing astonishing then if the results of the two sets of investigations do not agree altogether precisely. Nevertheless the agreement is very good. It shows that even my first process of extraction yields radiotellurium free from any other radioactive substance, however strongly it might be mixed with other, inactive substances.

In both investigations, the saturation current furnished the measure for the strength of the radioactivity. From the few lines comprising the preliminary publication of Meyer and von Schweidler, 135 days appears as the time in which the intensity of the radiation of the radiotellurium sinks to half-value, on the basis of the most reliable observations. "The decay follows approximately the law $e^{-\lambda t}$."

We have found likewise that the decay, as is to be expected for a homogeneous radioactive substance, follows the formula for monomolecular reactions. The following table shows how far the values calculated from the formula

$$\frac{J_t}{J_0} = e^{-\lambda t}$$

agree with our observations. The value of λ, taking into account all the observations, was calculated at $\lambda = 0.004959$ if t is expressed in days.

	$J_t : J_0$	
Number of days	Calc.	Found
70	0.707	0.725
97	0.608	0.591
128	0.530	0.514
260	0.275	0.265
319	0.206	0.210

Hence the intensity sinks to half-value in 139.8 days and the mean lifetime of the radiotellurium atom amounts to 201.7 days.

The constant of radioactivity is the most characteristic index of a radioactive substance. The value stated here will not undergo any important alteration from further investigations which are already in progress. Hence it is demonstrated that radiotellurium is a homogeneous radioactive material, and, contrary to the conjectures often expressed, is not identical with "polonium." Without doubt the Curies' polonium contained radiotellurium, as I have pointed out in the case of Giesel's polonium. However "polonium" is undoubtedly a mixture of radioactive substances. In her celebrated Dissertation, Madame Curie makes the following statement concerning the decay of her polonium: "A specimen of the nitrate lost half of its activity in 11 months and 95 per cent in 33 months. Other specimens behaved similarly. A specimen of the metal lost 67 per cent of its activity during 6 months." The first two numbers show that this polonium must contain more than one radioactive constituent, since the decay does not follow the formula for monomolecular reactions. The last number quoted shows in contrast to the first two a more rapid decay than radiotellurium, which loses 67 per cent of its activity in $7\frac{1}{2}$ months.

The relatively stable transformation product of radium discovered by Rutherford,[7] radium E whose activity sinks to half-value in a year, is not identical with radiotellurium as the investigation of Meyer and von Schweidler as well as our own shows. On the other hand, the decay constant agrees in a remarkable way with one of the values quoted above, which Madame Curie has repeatedly observed with her bismuth-polonium nitrate.

[When Marckwald announced the half-life of radiotellurium in January, 1905, Rutherford had already begun to suspect that there was a sixth substance in the chain of transformation products from radium emanation. By May he was sure.[8] It became radium D, a long-lived substance with no detectable rays. Radium E was the name he now assigned to a beta-emitter with a half-life of six days, and the final substance in the chain was now called radium F. It gave alpha rays only, dissolved in acids and deposited on a plate of bismuth dipped into its solution. Its half-life was 143 days. In all these properties it strongly resembled radiotellurium, whose half-life Meyer and von Schweidler had

---

[7] *Philos. mag. 8*, 636 (1904); *Philos. transact of the royal soc. 204*, 202 (1904).

[8] E. Rutherford, "Slow Transformation Products of Radium," *Nature, Lond.*, 1904–05, *71*: 341–342, *Phil. Mag.* [6], 1905, *10*: 290–306.

set at 135 days and Marckwald's young physicists at 139. Rutherford had no doubt that the two substances were the same, and that the radiotellurium in the Joachimsthal residues was there as a transmutation product of radium. Whether the Curies' polonium was also to be identified with radium F was a question he did not attempt to answer.

The answer came early in 1906 from Marie Curie, as set down in the next paper. She appears to have begun the work soon after Marckwald had published his value of the half-life, and it is interesting to notice how strongly she affirms the value of such measurements. This paper was presented in her name to the Académie des Sciences by Pierre Curie, who had been elected to its membership the previous July.[9]—A. R.]

10

Mme. Curie

On the Diminution of the Radioactivity of Polonium with Time

[Translation of "Sur la diminution de la radioactivité du polonium avec le temps," *Comptes rendus de l'Académie des Sciences, Paris*, 1906, *142*: 273–276 (29 January).]

About ten months ago I undertook a series of measurements with a view of determining the diminution of activity of polonium with time.

The polonium which served for this study was prepared according to the method indicated in the first publication relating to its discovery[10] and described in more detail in my doctoral thesis. It is necessary first to extract from the mineral the salt of bismuth which it contains. For that, the hydrochloric acid solution of the mineral is precipitated with hydrogen sulfide, the sulfides are separated, washed and dissolved in dilute nitric acid, and the solution is precipitated with water. The mixture of sub-nitrates and oxides so obtained is treated with a boiling solution of soda which removes the lead, arsenic, and antimony. By repeating the various treatments indicated, one can obtain a very pure polonium-bearing oxide of bismuth. To concentrate the polonium, one dissolves this oxide in nitric acid and undertakes a series of fractional precipitations with water, the portions precipitated most easily being those in which the activity concentrates.

---

[9] *C. R. Acad. Sci., Paris*, 1905, *141*: 24 (Session of 3 July), 81 (Session of 10 July).

[10] Curie, *Comptes rendus*, July 1898.

A suitable quantity of polonium-bearing bismuth oxide of moderate activity (250 times more active than uranium) was placed in a very shallow circular cavity hollowed from the central part of a circular metallic disk. The oxide powder, which filled the cavity, thus occupied a definite area on the disk. The radioactive plate thus formed was carefully preserved and its radioactivity was measured at suitable intervals of time. The intensity of the radiation was evaluated by the saturation current produced in a parallel-plate condenser and the intensity of this current was measured by our usual method, by means of an electrometer and piezoelectric quartz.

Here are the first results of these measurements: the intensity of the radiation diminishes as a function of the time according to a simple exponential law. Designating by I_0 the initial intensity, by I the intensity at time t, and by a a constant, one finds that

$$I = I_0 e^{-at} \qquad [I]$$

If t is expressed in days, $a = 0.00495$; according to this relation the intensity of the radiation diminishes to half its value in a time equal to 140 days. The deviations between this law and the measurements do not exceed 3 per cent.

The graphical representation of the results is given by Fig. 1. The curve V on this figure was obtained by plotting time as abscissa and log I as ordinate. This curve is a straight line to the approximation which had just been indicated.

It should be noticed that the constants which define the radioactive properties of substances play a role absolutely comparable with the wave lengths of lines in the spectra of the elements. When a radioactive material is mixed in very small quantity with an inactive material, the constants deduced from the study of its radiation serve to characterize it without ambiguity. The constant a of formula (I) then is characteristic of polonium.

The time constant which I have found for polonium affords certain proof that the substance studied by Marckwald under the name of *radiotellurium* is identical with polonium. This identity seemed almost evident according to all of Marckwald's publications on the properties of *radiotellurium*. It is made certain by the fact that the constant I have found for polonium is quite the same as that which Marckwald recently determined for his *radiotellurium*. Indeed Marckwald found $a = 0.00497$ as the value of a in the formula.[11]

Polonium and radiotellurium, then, are one and the same substance, and it is obviously the name of *polonium* which should be used,

---

[11] Marckwald, *Jahrbuch der Radioaktivität*, July 1905.

polonium being not only anterior to radiotellurium, but being the first strongly radioactive substance discovered by M. Curie and myself by a new method of investigation.

I have also used the method of concentration employed by Marckwald; this method is very convenient. It consists in dipping a

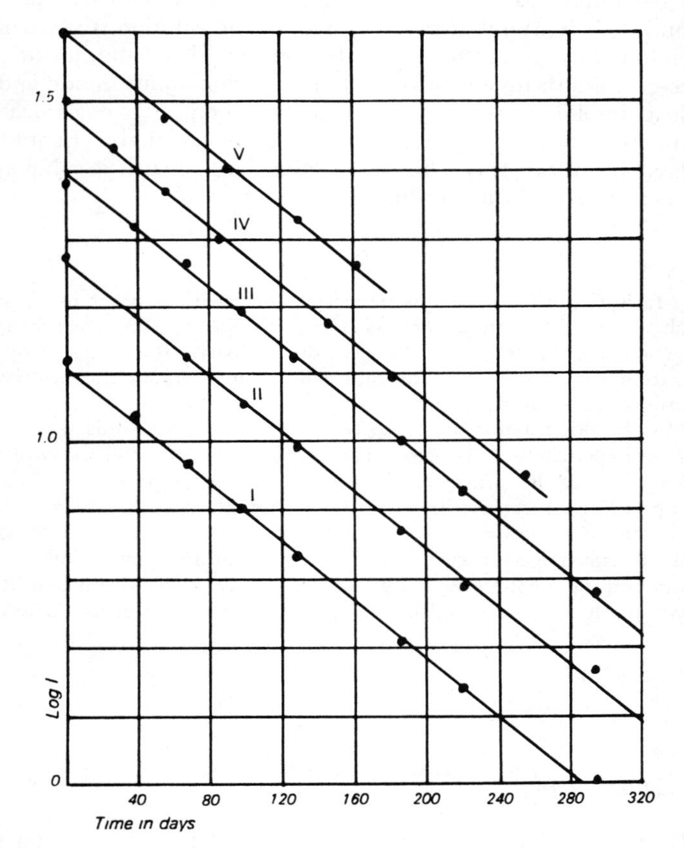

Fig. 1.

plate of bismuth into a hydrochloric acid solution of the salt of the radioactive bismuth. The polonium is deposited on this plate. I have used plates of platinum covered with a thin layer of bismuth by electrolysis, and I have concentrated polonium on these plates. They are very convenient for the study of radioactivity since they do not require the same precautions as plates covered with an oxide powder.

The lines *I*, *II*, *IV* of the figure were obtained with such plates. The line *III* was obtained from an active plate wrapped in aluminum leaf 0.01 mm. in thickness; it indicates the drop with time of the rays which penetrate that thickness of aluminum. All the lines are parallel.

It should be remarked in closing that it can in no way be claimed that polonium has the reactions of tellurium rather than those of bismuth. It is true that in hydrochloric acid solution it is partially precipitated by tin protochloride, but on the other hand, in the presence of bismuth its sulfide is insoluble in ammonium sulfide and its oxide is insoluble in a solution of boiling soda. To recognize the reactions of polonium we must have it in weighable quantities. Today we can say only what its reactions are when it is submerged in other materials which ordinarily accompany it.

[Marie Curie followed this paper instantly with another in German which she published in the *Physikalische Zeitschrift*.[12] Here she rehearsed the whole controversy with Marckwald, repeated the account of her measurements of the time constant, and firmly stated the priority of polonium over radiotellurium.

Marckwald did not have the temperament to take this quietly, as the next paper shows. However his claims are honest, and the capitulation at the end is complete. The quotation from *Romeo and Juliet* is perhaps a shade more picturesque in the original, where the English lines stand out from the German text. It is interesting that although both Marie Curie and Marckwald had accepted the transformation theory, neither considered the possibility that radium E (Bi^{210} with a half-life of 6 days) might be present initially in polonium preparations.—A. R.]

11

W. Marckwald

On Polonium and Radiotellurium

[Translation of "Über Polonium und Radiotellur," *Physikalische Zeitschrift*, 1906, *7*: 369–370. (1 June).]

In a communication by Mme. Curie on the time constant of polonium which has recently appeared in this journal,[13] it appears

---

[12] Frau Sklodowska Curie, "Über die Zeitkonstante des Poloniums," *Phys. Z.*, 1906, *7*: 146–148 (1 March).

[13] This Journal *7*, 146, 1906. Cf. *Comptes rendus 142*, 273, 1906.

that the author has now succeeded in obtaining, by fractional precipitation from the bismuth subnitrate obtained from pitchblende, a product whose radioactivity decays by the same law as that of radiotellurium.[14] I have demonstrated in other places[15] that the "polonium" which was previously described contained a mixture of radioactive constitutents, a conclusion supported by Mme. Curie's own statements on the decrease in the radioactivity of polonium. Mme. Curie will not grant that this argument applies, since she began her earlier observations with no intention of determining the decay constant and without any special precautions.

The question whether Mme. Curie's "polonium" contains a homogeneous, radioactive substance can hardly be settled except by her own data. It is not to be contested that these data are not to be used for conclusions which their author did not have in view. Mme. Curie has never given the data on the decrease in the activity of her polonium[16] in a form which would warrant the assumption that her observations were not exact. But even within this limitation, no doubt can arise that the "polonium" contains another radioactive substance alongside the radiotellurium. The deviations in the decrease in its radioactivity from the law which holds, as I have shown, for radiotellurium, amount to 100 per cent, and this is in no way to be explained by uncertainties of observation.

When I began my investigations on the radioactive bismuth from pitchblende,[17] four years had passed since the first publication of P. and S. Curie on polonium.[18] In this time there had been no important advance in the chemical investigation of the substance. The discoverers had succeeded by a few fractionations in increasing the radioactivity of polonium, which exceeded that of uranium a hundred times, to the level of four hundred times. When I found the first process for separating the radioactive constituent quantitatively from the bismuth by a chemical reaction, the deposited material appeared in no way homogeneous. Rather it consisted for the greater part of tellurium. By suitable reagents, the radioactive constituent could be smoothly separated from the tellurium. Then for the first and the only time so far a highly radioactive substance was brought out which could be recognized by characteristic chemical reactions as a new

---

[14] Meyer and v. Schweidler, *Wiener Anzeiger* 1 Dec. 1904. Marckwald, *Ber. d. deutsch. chem. Ges. 38*, 591, 1905.

[15] *Jahrb. d. Radioaktivität u. Elektronik 2*, 133, 1905

[16] *Untersuchungen über die radioaktiven Substanzen,* trans. by Kaufmann, Braunschweig 1904, p. 110.

[17] *Ber. d. deutsch. chem. Ges. 35*, 2285, 1902.

[18] *Comptes rendus 127*, 175, 1898.

element and separated from all known elements. For radium, as is known, few such reactions have been found, as also for actinium, radiolead, etc. How Mme. Curie can write with regard to these facts, "On the whole we can say nothing certain at present concerning the chemical properties of pure polonium," yet still identify polonium and radiotellurium, is incomprehensible to me. The chemical properties of radiotellurium have been so precisely determined, that the almost quantitative separation of the few milligrams of that substance which were contained in 5000 kilos of uranium residues, was carried out without difficulty by the chemical reactions which I have found.[19] That "chemically pure" radiotellurium has not yet been obtained is not a consequence of its chemical properties or of any deficiency in method, but simply of a deficiency in material.

Radiotellurium makes up by weight hardly the ten-millionth part of the bismuth from pitchblende. It is therefore several million times more radioactive than polonium is described to be before its separation, and surpasses by some thousand-fold the strongest polonium preparation which Mme. Curie had been able to concentrate by her laborious fractionations before my publication.

I thought that I should give the readers of this journal this short review of the history of polonium so that they might form their own judgment whether the following sentence from the previously cited work of Mme. Curie's is justified: "· · · there can be no question, therefore, that polonium contains radiotellurium as Marckwald indicates; however, the substance prepared by Marckwald is simply the same as I discovered earlier and have described as polonium."

At the close of her paper Mme. Curie disputes that polonium (radiotellurium) stands nearer tellurium than bismuth as regards its chemistry. I named the substance radiotellurium "provisionally" since all its chemical properties required it to take that place in the sixth group in the periodic system for an element of a somewhat higher atomic weight than bismuth which is still unoccupied. Such an element would be expected to be more negative than bismuth, more positive than tellurium, and more difficult to reduce from its oxide than the latter element; further its oxide should have basic rather than acidic properties. All of this corresponds to radiotellurium. Rutherford has further shown that radiotellurium is the last radioactive decay product of radium, and that it probably disintegrates into helium and lead. For this substance an atomic weight of around 210 would be expected.

---

[19] Marckwald, *Ber. d. deutsch. chem. Ges. 38*, 591, 1905.

On these grounds I hold the name of "radiotellurium" to be entirely suitable. However:

> What's in a name? that which we call a rose,
> By any other name would smell as sweet.

The great services of Mme. Curie in the discovery of the radioactive substances justify us in considering her wishes in a question of no wide-ranging importance. For this reason I propose in the future to replace the name of "radiotellurium" by "polonium."

Berlin, April 1906.

(Received 20 April 1906).

IV. Radiochemistry Uses the Transformation Theory

[In the paper which follows, Otto Hahn (1879–1968), a visitor in Ramsay's laboratory at University College, London, begins at the point which Marckwald and Marie Curie had reached in that same winter of 1905. That is to say, he characterizes his new substance, and indeed recognizes that it *is* new, by reference to its radioactive properties rather than its chemical reactions. He had little choice, since the chemical operations which could be performed on a 10-milligram or 20-milligram specimen were distinctly limited. Nevertheless the adoption of the new criteria of identity increased the possible range of investigations enormously. Substances could now be studied which were far too rare to yield an atomic weight or even a characteristic spectrum.

Hahn was nearly as ill-equipped to make this discovery as the Curies had been when they isolated polonium and radium. He was trained as an organic chemist and what was needed here was skill in inorganic chemistry, especially as it applied to the rare earths. Yet perhaps Ramsay was not altogether joking when he commended Hahn's lack of experience as a guarantee of freedom from prejudice.[1] He had the wit to look for an emanation, that is for the evolution of a radioactive gas, and when he found one, he learned how to distinguish the emanations of Giesel's emanium (which many took to be identical with Debierne's actinium) from the emanations of thorium and radium.

It is worth noticing how clearly Hahn could distinguish his new substance from thorium, even though its only characteristic properties were those which thorium was known to possess.—A. R.]

[1] O. Hahn, "Einige persönliche Erinnerungen aus der Geschichte der natürlichen Radioaktivität," *Die Naturwissenschaften*, 1948, *35*: 64–74.

12

O. Hahn

A New Radio-Active Element which Evolves Thorium Emanation. Preliminary Communication

[From *Proceedings of the Royal Society, A*, 1905, *76*: 115–117.]

(Communicated by Sir William Ramsay, K.C.B., F.R.S. Received March 7,—Read March 16, 1905.)

[The material for this investigation was provided by Sir William Ramsay; it was the final residue remaining after fusion with bisulphate of 5 cwt. of the cubical ore from Ceylon, for which the name "thorianite" has been suggested by Professor Dunstan. This residue was fused with carbonates, the silica was removed, and the carbonates dissolved in dilute hydrochloric acid. Lead was precipitated as sulphide, and the carbonates again precipitated. These preliminary operations were carried out by Mr. Charles Tyrer and by Dr. Denison.]*

This residue weighed about 18 grammes, and a preliminary estimation of radio-activity led to the belief that it would yield about 15 milligrammes of pure radium bromide. The carbonates were dissolved in pure aqueous hydrobromic acid, and the bromides fractionated according to Giesel's method. But difficulties were soon encountered; the more soluble portion did not fall off in radio-activity, but gradually grew more strongly radio-active; the radium concentrated at the least soluble end, and the middle fractions became relatively weak in radio-activity.

Small traces of iron and other impurities, unavoidable in London, collected in the more soluble portions, and the ferric bromide imparted to them a brownish-yellow colour. These, and indeed all fractions, were again treated with hydrogen sulphide, and a minute quantity of a peculiar dark-brown precipitate came down. It was also radio-active; it was soluble in nitric acid to a pale green solution, and on evaporation crystals of two kinds deposited; easily soluble green crystals and less soluble white ones. The investigation of these bodies is still in progress.

* [The square brackets here are Hahn's.—A.R.]

By a series of troublesome operations, a quantity of precipitate was obtained by aid of ammonia, and to separate iron, it was treated in acid solution with ammonium oxalate; this produced about 10 milligrammes of crystalline precipitate, which was by far the most active preparation obtained, and which shows after two months no diminution in its radio-active power. It glows feebly in the dark, and imparts bright luminosity to screens both of platino-cyanide and zinc sulphide. If a current of air be blown through a solution of this substance and directed on to a screen coated with zinc sulphide, luminosity is produced, which, nevertheless, is different in intensity from that shown when a similar experiment is performed with Giesel's emanium. The phenomena are not so brilliant as those obtained from a strong sample of emanium kindly sent by Professor Giesel. It was not possible to perform the beautiful experiment of allowing the emanation to pour down on the screen and blowing it away, probably because the new substance emits β-rays in too great abundance. But that the dry substance also evolves emanation was easily discovered by help of an electrometer.

The first impression, that the new substance was identical with actinium or emanium, was found to be untenable,[2] for the new preparation evolves an emanation identical with that of thorium; different samples gave for the half-period of decay from 52 to 55 seconds: for the half-period of the induced activity, somewhat more than $11\frac{1}{2}$ hours was found, and a small remaining activity persists and decays very slowly. (The half-period for thorium emanation was found by Le Rossignol and Gimingham[3] to be 51·2 seconds: Bronson, working in Rutherford's Laboratory, found 54 seconds.) As this phenomenon has up till now not been noticed with thorium emanation, it may be conjectured either that another radio-active substance is mixed with the new body in very small traces, of which the induced radio-activity must have a long period of decay, or what is less probable, that the induced activity of thorium, like that of radium, changes into another product with a long radio-active existence. It is certain that radium emanation and also Rutherford's radium-E were absent.

The oxalate, which weighed 10 milligrammes, dissolved in hydrochloric acid, gave a quantity of emanation considerably greater than would be evolved from a kilogramme of thorium in solution; consequently, it is more than 100,000 times as active as thorium. Further

---

[2] The measurements of the emanations and excited activities were carried out in collaboration with Dr. Sackur, working in this laboratory; we also redetermined the half-period of decay of the emanation from Giesel's emanium, as about three seconds, and for its induced activity a period of about 36 seconds. More exact measurements are in progress.

[3] 'Phil. Mag.,' July, 1904, p. 107.

work has resulted in the accumulation of 20 milligrammes of material nearly 250,000 times as active as its own weight of thorium nitrate. Thorium itself, if present at all, must be there in minimal quantity, for the oxalate gives tests for calcium for the most part. Whether this active substance is a constant radio-active constituent of thorium preparations, or whether it is another new radio-active element, remains still undecided. Its quantitative extraction from thorium salts has not yet been investigated. After precipitation of a small part of the solution in hydrochloric acid of the original ammonia precipitate with ammonia, the filtrate shows considerable radio-activity, which rapidly falls off in a few days, but does not wholly disappear, and the removal of this substance does not diminish appreciably the radio-activity of the insoluble residue. Whether that is due to thorium-*x* or not has not yet been quantitatively investigated. The close relation of the new body to thorium is proved, not merely by the apparent identity of the two emanations, but also in its having been separated from a mineral unusually rich in thorium.[4]

We are in hopes that it may prove possible, by several processes of concentration, to obtain an even more strongly radio-active product, and to be able to describe more in detail the chemical properties of the substance; one difficulty consists in the adhesion of the substance to all precipitates; all filters are radio-active, and can hardly be purified by repeated washing. The activity of the sulphide precipitate may, perhaps, be due to this cause; the emanation which it yields appears to be identical with that obtained from the precipitate with ammonia.

Recent researches would appear to show that this substance is present in soil in amount comparable with, but still considerably smaller than radium. G. A. Blanc[5] has described in a paper on "the Radio-Activity of Mineral Springs," a gas which contains thorium emanation. N. M. Dadourian,[6] in investigating the radio-activity of subterranean air, has detected not only radium emanation, but also that of thorium; and Elster and Geitel[7] have described a preparation obtained from the mud from the Baden Baden "Ursprung" as containing no thorium in detectable quantity, but yet evolving thorium emanation in amount such that half a gramme of thorium oxide would be required to produce it. They conjecture, therefore, the presence of a new radio-active element. Attention may also be called to

---

[4] Experiments are in progress in this laboratory to attempt to concentrate the radioactive substance from a large quantity of thorium; but, so far, no definite results have been obtained.

[5] 'Phil. Mag.,' vol. 9, pp. 148 to 154.

[6] 'Sill. Amer. J. of Science,' vol. 19, 1905, pp. 16 to 22.

[7] "Radioaktivität der Sedimente der Thermalquellen," 'Chem. Centralbl.,' 1905, vol. 1, p. 651.

the fact that inactive thoria is said more than once to have been obtained.

It is almost certain that all these emanations are the product of this new substance, and are not derived from thorium itself, for the amount of emanation obtainable from thorium is so small that, if it can be measured at all, it should be possible to detect thorium analytically in the source from which it is evolved.

[Hahn's final reports on his new radioactive substance[8] added little more than detail to the preliminary announcement. He did determine that it produced the same thorium X which Rutherford and Soddy had established as the precursor to the emanation. In addition, for its resemblance to thorium on a deeply intensified scale of activity, he named it radiothorium.

Ramsay was thoroughly impressed and proceeded to secure Hahn an appointment in Emil Fischer's laboratory at the University of Berlin.[1] Hahn, with a keener appreciation of his own ignorance, arranged to postpone the appointment for a year and spent the winter of 1905–06 with Rutherford at McGill, learning the techniques of research in radioactivity and mastering the mathematics of the transformation theory. By the time he reached Berlin, the existence of radiothorium had been thoroughly confirmed by others.[9]

It had also been established that radiothorium was a transformation product of thorium, or in the genetic language which was often used, that it was descended from thorium. In radioactivity, the rate at which a substance transforms itself is taken to be proportional to the quantity of that substance present, that is

$$\frac{dQ}{dt} = -\lambda Q.$$

If this substance is formed by the transformation of some other, then the rate of disappearance of the predecessor is simply the rate of growth of the succeeding substance, or

$$\frac{dQ}{dt} = \lambda_1 P - \lambda_2 Q.$$

[8] O. Hahn, "Über ein neues, die Emanation des Thoriums gebendes radioaktives Element," *Jb. Radioakt.*, 1905, *2*: 233–266; "Ein neues radioaktives Element, das Thoriumemanation aussendet," *Ber. dtsch chem. Ges.*, 1905, *38*: 3371–3375.

[9] F. Giesel, "Ueber die 'Thor-aktivität' des Monazits," *Ber. dtsch chem. Ges.*, 1905, *38*: 2334–2336.

G. A. Blanc, "Über die Natur der radioaktiven Elemente, welche in den Sedimenten der Thermalquellen von Echaillon und von Salins-Moutiers (Savoyen) enthalten sind," *Phys. Z.*, 1905, *6*: 703–707; "Untersuchungen über ein neues Element mit den radioaktiven Eigenschaften des Thors," *Phys. Z.*, 1906, *7*: 620—630.

J. Elster and H. Geitel, "Beiträge zur Kenntniss der Radioaktivität des Thoriums," *Phys. Z.*, 1906, *7*: 445–452.

When dQ/dt vanishes, the material reaches a state of *radioactive equilibrium*, and then

$$\frac{P}{Q} = \frac{\lambda_2}{\lambda_1},$$

that is to say, the quantity of each substance present is inversely proportional to its transformation constant λ. In the radioactive minerals, equilibrium is completely established, and one can show by an extension of this line of reasoning that every descendant of the original uranium or thorium is present in a constant ratio of abundance to the uranium or thorium content. Working by this theory, McCoy and Ross at Chicago, Boltwood in his own laboratory in New Haven, and Dadourian at Yale had all shown that the radioactivity of radiothorium in various minerals was strictly proportional to their thorium content, and thus that radiothorium must be a transformation product of thorium. They found also a constant ratio, but a smaller one, for commercial thorium salts which indicated some removal of radiothorium during their preparation.[10]

The modern reader, who can identify radiothorium as Th^{228} and the new intermediate product as Ra^{228}, can appreciate the difficulties which Hahn resolved in the next paper. He had differentiated radiothorium from thorium, yet any extraction of thorium from its minerals brought with it the equilibrium quantity of radiothorium. On the other hand, any commercial thorium salt gave evidence that the original radiothorium had been removed during its processing. The argument which established the new substance, *mesothorium*, as the separable material can be followed here, and it is to be noticed how much it depends on physical reasoning based on the growth and decay of the radioactivity of succession products. The reasoning is correct and convincing, but one may guess that chemical information on the nature of mesothorium was decently suppressed in deference to Knöfler's commercial interest.—A. R.]

---

[10] B. B. Boltwood, "The Radio-Activity of Thorium Minerals and Salts," *Amer. J. Sci.* [4], 1906, *21*: 415–426.

H. M. Dadourian, "The Radio-Activity of Thorium," *Amer. J. Sci.* [4], 1906, *21*: 427–432.

H. N. McCoy and W. H. Ross, "The Relation Between the Radio-Activity and the Composition of Thorium Compounds," *Amer. J. Sci.* [4], 1906, *21*: 433–443.

13

O. Hahn

A New Intermediate Product in Thorium

[Translation of "Ein neues Zwischenprodukt im Thorium," *Physikalische Zeitschrift*, 1907, *8*: 277–281 (1 May).[11]]

Recently a whole series of works has appeared on the properties of radiothorium and in particular its position with regard to thorium. In my first comprehensive paper on radiothorium[12] I emphasized the probability that it is a disintegration product of thorium, that thorium itself emits no rays, and that the activity of ordinary thorium arises only from radiothorium and its disintegration products. It was to indicate this that I chose the name of radiothorium for the highly active product.

The works of G. A. Blanc[13] and of Elster and Geitel[14] confirmed and improved our knowledge of the radioactive properties of radiothorium and their correspondence with those of thorium, so that no doubt prevails concerning the identity of the pair of emanation-emitting substances.

Elster and Geitel succeeded in preparing from ordinary thorium a small quantity of an oxide which kept a constant activity about twelve times greater than that of ordinary thorium oxide. They have also attained a certain concentration of radiothorium. Also G. A. Blanc through the well-known action of barium sulfate was able to obtain from 6 kg of thorium nitrate a few milligrams of a product which was several thousand times stronger than thorium oxide and also maintained its activity.

The papers of Boltwood[15] and Dadourian[16] report on the relation of the activity of thorium minerals to their content of thorium; they lead independently to completely concordant results. Both investigators found that the activity of thorium minerals selected for equal

---

[11] [A parallel account with the same title was submitted simultaneously to the *Deutsche Chemische Gesellschaft* and was published in *Ber. dtsch chem. Ges.*, 1907, *40*: 1462–1469.—A. R.]

[12] Jahrb. d. Rad. u. Elektr. *2*, 233, 1905.

[13] This journal *7*, 620, 1906.

[14] This journal *7*, 445, 1906.

[15] This journal *7*, 482, 1906.

[16] This journal *7*, 453, 1906.

thorium content is a constant number; a result which proves that radiothorium is actually a disintegration product of thorium.

There exist also similar relationships for uranium and radium; thus their genetic relationship can also be concluded from the constancy of the ratio of uranium to radium in the most varied minerals.

The investigations of Boltwood and of Dadourian were not limited however to thorium minerals; commercial preparations of thorium were also drawn into the circle of their investigations. Thus they found the very striking result that commercial thorium nitrate shows an activity of less than half the value it ought to exhibit from its thorium oxide content. For this the two experimenters used entirely different methods of measurement. Boltwood compared the activity by the action of the alpha rays in an experimental procedure like that first devised by McCoy.[17] Dadourian converted his preparations into the form of solutions, collected the active deposit originating from the emanation on a negatively charged wire, and measured the activity so acquired under constant external conditions. Most recently Eve[18] has come to the same results by comparing thorium minerals and commercial preparations of thorium of known content with the help of their gamma rays.

Commercial thorium nitrate was always less than half as active as it should have been.

Boltwood concluded from these results that in the technical manufacturing process of thorium nitrate more than half of the radiothorium was separated from the thorium.

Such a possibility was not indicated at first glance, indeed such a thorough separation is quite remarkable when one considers that up to now many efforts at the separation of radiothorium from thorium have been either completely fruitless, or, as in the case of the experiments of Elster and Geitel and of G. A. Blanc, have proceeded only so far that a difference in intensity from the proper activity of thorium could barely be established.

I began a systematic investigation of the activity of commercial thorium preparations of various grades of purity to test whether the processes of purification permitted any decrease in activity to be established. These experiments were made possible by the obliging cooperation of the firm of Dr. O. Knöfler and Co. in Plotzensee bei Berlin, who placed at my disposal a variety of assay samples, for which I here offer the firm my heartiest thanks.

The activity of the preparations was measured by McCoy's procedure,[17] which made it possible to lay down the samples in a layer

---

[17] Journ. Americ. Chem. Soc. *27*, 291; Am. Journ. Science *21*, 433.
[18] Am. Journ. Science, Dec. 1906.

of the desired thinness and evenness of distribution, so that the various preparations could be compared directly by their alpha activity under otherwise uniform conditions. In general the procedure was to convert each thorium preparation into the oxide, which was ignited for a short time to bring it into the non-emanating condition. Then it was mixed with chloroform, painted on a plate of aluminum with a fine brush, and measured in the electroscope. The plates used had the same shape and area; the quantity of substance tested amounted to 10–12 mg in general. Preliminary tests showed that for equal preparations under these conditions the values found were in satisfactory agreement.

From time to time there came preparations to be measured whose activity indicated a high content of thorium but which had still undergone repeated purifications. When the somewhat varying composition of the oxides was taken into account, the general result was that during the process of purification no overall or sudden weakening of the activity could be observed for certain. Variations in activity appeared indeed, but rather in the contrary sense, since the less pure specimens were occasionally found somewhat weaker than the pure, calculated for the same content of thorium. The basis probably lay in a small quantity of alkali which made the layer somewhat hygroscopic and produced a somewhat stronger absorption of the alpha particles. (It is to be remarked that thorium oxide which is not sufficiently ignited is also hygroscopic.)

The separation of the radiothorium, if it occurs at all, must occur during the preparation of the raw thorium from the monazite. I therefore tested a few preparations and waste liquors from the first processes in the treatment of the monazite so far as they could be placed at my disposal. However here the result was the same. The activity answered in satisfactory style to the approximate content of thorium, and even a working-up of the insoluble and undissolved residues showed, after an excess of thorium X had disintegrated, no considerable increase in thorium activity.

These investigations to find a point in the production and purification of commercial thorium at which the separation of radiothorium takes place or to detect a general weakening during the continued processes of purification came to nothing and the question of the location of the missing radiothorium remained open.

I was strengthened then in a conjecture which I had entertained earlier, that perhaps the disintegration of thorium into radiothorium does not proceed directly, but that an intermediate product may exist from which the radiothorium is formed.

Various phenomena seemed to speak for the existence of such a substance:

According to a private communication from Boltwood, his weakly active thorium nitrate showed no noticeable increase in activity over the course of almost two years. This would lead to the conclusion that radiothorium has a relatively long life. On the other hand I have already indicated in an earlier communication that the radiothorium preparations which I have investigated show a distinct decrease in their activity. In my last communication on the subject I mentioned that a series of precise measurements was in progress which, avoiding errors as far as possible, would clear up the disintegration period of radiothorium. To eliminate outside influences like dampness, etc., the preparation was stored in an air-tight glass dish under a thin sheet of mica. Unfortunately this preparation was destroyed on the journey from America so that a new experiment had to be set up again here. A preparation sealed into a glass tube was tested by its beta activity. Another, similar to the preparation mentioned, was put in a watchglass and covered air-tight with a thin sheet of mica. To avoid electrostatic disturbances the containers were completely wrapped in tinfoil. The result once again is that the activity of the preparations decreases. However I cannot yet give a definite disintegration constant since the decrease recently has been much slower that it was earlier. For order of magnitude, two years appears to me an upper limit for the disintegration period of radiothorium; perhaps it should be shorter. I shall return shortly to a possible explanation why no single value has been found for the decay.

Radiothorium preparations thus decrease, yet Boltwood's weakened thorium does not increase, or not in the same measure as one would expect if radiothorium were the direct descendant of thorium. At this point there is a way out of the difficulty if one assumes the existence of an intermediate product between thorium and radiothorium with a longer life than that of radiothorium. This would explain why the weak thorium preparation mentioned does not increase in the same measure that the radiothorium preparations I have investigated decrease. One need only assume that the thorium preparation lacks not only radiothorium but also the intermediate product whose rebuilding must precede the rebuilding of the radiothorium.

Also the presence of a considerable thorium activity in the water-laid sediments of G. A. Blanc and of Elster and Geitel would be easier to explain if one assumed a slower product whose disintegration constant amounted to less than 2 years.

To prove the possible existence of such an intermediate product between thorium and radiothorium, the best way would be to compare thorium preparations of the same strength and method of manufacture but of different ages, to test whether differences in activity could be detected. Here I am fortunate in the support of the firm of O. Knöfler and Co., who placed at my disposal average samples of pure thorium nitrate from different years, including a few whose date of preparation was exactly known.

Their investigation disclosed in fact quite considerable differences in activity, specifically that a gradual lowering of activity could be detected from the newest to the older samples. Freshly prepared thorium nitrate showed an activity of the same order of magnitude as that of a corresponding quantity of thorianite when the activity arising from the uranium and radium in the mineral is deducted. This result is identical with Boltwood's finding when he prepared his own thorium nitrate, whose strength was of the right order of magnitude.

Next, over a few years, the activity decreases and after about 3 years appears to reach a value at which it remains almost constant for a considerable period of time. Naturally I did not have the most desirable choice of well-defined preparations at my disposal; most of the older products were average samples for the year, which gave no truly exact values. For the time being I am not able to give a uniform curve of the activities found for any great period of time; neither is it certain here whether in each case exactly the same conditions prevailed during the manufacture. It is obvious that the gradual decrease in activity of the freshly prepared thorium samples arises from the disintegration of an excess of radiothorium which is present for some reason; indeed it gives as an order of magnitude for the decrease a period of about two years.

After the lowest value of the activity has been reached and has remained practically constant for a time, a gradual upward turn seems to appear. A few very old preparations, one in particular from the year 1898 and one very old sample from February 1895 display stronger activities; indeed the preparation from 1895 shows a somewhat higher value than the one 3 years younger.

In this communication I shall dispense with the reproduction of any numerical results since I hope shortly to report on a greater selection of old thorium preparations.

I shall content myself here with the interpretation of the values I have found.

In the technical manufacture of thorium nitrate an intermediate product is separated which emits no alpha particles. The radiothorium itself remains with the thorium. Thus the activity of the nitrate immediately after its manufacture is of the expected order of magnitude. The disintegration time, and therefore the reconstitution time, of the intermediate substance is longer than that of the radiothorium. The activity of the fresh preparation therefore declines since the generator of the radiothorium is missing, and does so with the period of the radiothorium. If the intermediate product were entirely separated and if it had a very long life, the decrease would go almost to zero. As a matter of fact the lifetime of the intermediate substance is not very great. Its disintegration period to a first approximation appears to be some seven years.

Then gradually the intermediate product will be formed again and with it the radiothorium. After a few years a minimum activity will be reached, and a gradual upward turn appears. The recovery of the thorium proceeds then with the slower period of the intermediate substance; when this is completely reconstituted the activity is again as high as immediately after the manufacture and remains constant. To attain this should require some 40 years.

From the gradual decrease in the activity of the thorium preparations, the existence of an intermediate substance can be proved indirectly.

A positive proof will be obtained when it is possible to produce from old thorium preparations or some product of the manufacturing process, thorium-free substances which at first show no activity or a negligible one and in which a gradual appearance of the specific activity of thorium can be recognized. The characteristic emanation of thorium is a sure guide here and excludes any error.

I have set up a large number of experiments in this direction *and I have succeeded in producing direct evidence of the new formation of radiothorium.* I have three preparations which contain no appreciable quantity of thorium and which during two months have shown a constant increase in activity. Another small specimen was prepared in November of last year and no slowing down in its increase of activity can be noticed. In the case of the latter only the action of the alpha particles has been studied, but of the newer preparations a part remains in solution and for this solution an increase in emanating power in constant amount can be detected.

Since in a preparation of radiothorium, even if initially it is completely free from thorium X, no change in emanating power occurs after a few months when it is tested under identical conditions, *this constant increase furnishes direct proof for the presence of a generator of radiothorium which is different from thorium.*

With the existence of this intermediate substance the experimental results of the investigators mentioned above can be explained without contrivance. Obviously they had at their disposal preparations which were already several years old and in which therefore the original excess of radiothorium had disintegrated entirely or for the most part. Boltwood in the communication cited above gives for the activity of various commercial thorium preparations values which differ considerably from one another in contrast to the remarkable constancy of activity of the minerals, and the differences lie well outside the limits of error. The explanation lies in the difference in age of the preparations obtained from various sources.

The fact that Boltwood had a weak preparation of thorium which showed hardly any change in its activity during almost two years appears to mean that this sample already lies at the minimum of its activity; on the assumption of an intermediate product, as stated above, it should now be possible to establish a slow increase in strength.

Also the disagreements and the apparent gradual slowing down of the decay times of the radiothorium preparations I have studied can find an explanation in the varying content of the intermediate product. Naturally it would have to be assumed that the radiothorium I prepared at the start from thorianite was not actually pure radiothorium but was mixed with a considerable quantity of the intermediate substance.

I shall postpone the final choice of a name for the new substance until its nature can be established more exactly. In the end, the name of "mesothorium" would seem appropriate.

Berlin, Chemical Institute of the University, 22 March 1907.

(Received 26 March 1907.)

[Bertram B. Boltwood (1870–1927), the author of the next paper, was an extraordinarily skilful analytical chemist who had been in private practice in New Haven. He had become interested in radioactivity, in particular in establishing by equilibrium methods the genetic relationship of radium to uranium and of radiothorium to thorium. This work had brought him into close scientific friendship with Rutherford and had led to his appointment in the fall of 1906 as Assistant Professor of Physics at Yale.[19] He knew Hahn through Rutherford and they had corresponded on the problems of radiothorium, as both their papers testify. Much of what Boltwood reports here on the radiothorium activities of various thorium oxides of American origin is a confirmation of Hahn's discoveries.

The paper is more interesting however as a first step toward the discovery of isotopes. By a shrewd intuition, Boltwood recognized the chemical similarity of radiothorium and thorium as characteristic for these substances and so guessed that mesothorium would resemble

---

[19] See the biography by Alois F. Kovarik in *Dictionary of American Biography*, ed. Allen Johnson, New York: Charles Scribner's Sons, 1929, Vol. 2, pp. 425–426; also A. S. Eve, *Rutherford*, New York: Macmillan, 1939, Chap. 4–5.

the other known member of that genetic chain, Rutherford and Soddy's thorium X. Since he was dealing with isotopic pairs, Th^{232} and Th^{238} for one set, Ra^{228} and Ra^{224} for the other, he found the likeness to be very close indeed. Just what he understood by this "similarity in chemical behaviour," is now impossible to guess, whether he saw only a rare-earth resemblance or perhaps glimpsed the chemical identity which Soddy would proclaim four years later.—A. R.]

14

Bertram B. Boltwood

On the Radio-Activity of Thorium Salts

[From *The American Journal of Science* [4], 1907, *24*: 93–100 (August).]

Measurements of the α-ray activities of a number of minerals containing thorium have been described in an earlier paper.[20] A measure of the ionization produced by known weights of the finely-powdered minerals in the form of thin films was obtained by introducing the films into an electroscope and determining the rate of leak of the charge in terms of the fall of the gold-leaf in scale divisions per minute. On dividing the rate of leak by the weight in grams of the mineral in the film a number was obtained expressing the specific activity (activity per gram) of the given mineral. The minerals examined contained uranium as well as thorium. The activity of one gram of uranium with its equilibrium amounts of disintegration products (actinium, radium, etc.) has been found to be a constant[21] which will be called the normal specific activity of the uranium-radium series. The value of this constant for the particular electroscope used was determined by measurements of the activity of certain minerals containing uranium and no thorium. Knowing the content of uranium and the constant for the normal specific activity of the uranium-radium series, it was possible to calculate for each of the thorium minerals that portion of the specific activity of the mineral which was due to the thorium and thorium products which it contained. Dividing this by the weight (in grams) of thorium contained in one gram of mineral, a number representing the activity per gram of thorium was obtained. It was found that, within the limits of experimental error, the activity corresponding to one gram of

---

[20] This Journal, xxi, 415, 1906.

[21] McCoy, Phil. Mag., ix, 176, 1906; Boltwood, loc. cit.

thorium in a mineral was the same for the different minerals examined, which indicated that the activity of one gram of thorium with its equilibrium amounts of disintegration products—the normal specific activity of the thorium series, as it can be called—was a constant.

Measurements were also made of the activities of a number of specimens of thorium oxide which had been separated by chemical methods from the minerals and from certain thorium salts prepared by the Welsbach Company. The relative ionization produced by a known weight of each of these oxides in the form of a thin film was determined in the electroscope. The specific activity of the oxides prepared directly from the minerals was found to correspond to the normal specific activity of the thorium series found in the minerals. The activity of the oxides prepared from the Welsbach salt was, however, found to be much lower than the normal. These results led to the conclusion that the chemical process employed by the Welsbach Company was in some way peculiar since it apparently resulted in the separation of over one-half of the radiothorium corresponding to the thorium present.

The question of the radio-activity of thorium in minerals and salts has been examined also by Dr. Dadourian,[22] who employed a method based upon the measurement of the activity of the deposit obtained by exposing a negatively charged plate to the emanation evolved by a solution of the thorium salt or mineral. The results obtained by Dadourian and the writer were in close agreement and led to similar conclusions. Results of a similar character were also obtained by McCoy and Ross[23] and by Eve.[24]

Among the salts examined by Dadourian were two specimens of thorium nitrate, the one prepared from North Carolina monazite and the other from Brazilian monazite, which had been supplied by the Welsbach Company to the writer nearly two years before. Mr. H. S. Miner, the chemist of the Welsbach Company, stated that the former salt was about two years old and the latter at least one year and a half old at the time they were sent to me. A third salt examined by Dadourian was a specimen of thorium nitrate which had been purchased from Eimer & Amend about three years before the time at which he tested it. It had, however, been used in the meantime by the writer for the preparation of thorium-X, that is, the original salt had been dissolved in water, the thorium had been precipitated as hydroxide with ammonia, and the washed hydroxide had been reconverted into nitrate. In order to again obtain the solid salt the solution

---

[22] This Journal, xxii, 427, 1906.
[23] Ibid., xxi, 433, 1906.
[24] Ibid., xxii, 477, 1906.

of the nitrate had been evaporated to dryness under conditions identical with those under which a considerable number of other salts have been prepared and which give a salt containing about forty-eight per cent of thorium oxide. The number given by Dadourian as expressing the specific activity of the thoria in this salt is therefore undoubtedly too high and the correct value would be about 9·0. The reason for this low value will appear later in this paper.

The specimen of Welsbach salt examined by the writer consisted of a kilogram of thorium nitrate made from North Carolina monazite and had been received about fifteen months before the time of my experiments.

The three Welsbach salts examined by Dadourian and the writer were therefore at least four years old, three and one-half years old and one and one-third years old, respectively. The oldest salt contained about forty per cent of the radiothorium in equilibrium with the thorium present and the youngest salt must have contained at least thirty per cent of its equilibrium amount of radiothorium. The difference in ages of these two salts was about three years and they had both been prepared from the same mineral by the same process. If radiothorium was a product formed directly from thorium it was obvious that its period of decay (recovery) could not be less than half-value in about six years and might be somewhat longer.

In April, 1906, the belief that the above data might have an important bearing in indicating the rate of disintegration of radiothorium was privately communicated to Dr. Otto Hahn, the discoverer of radiothorium. Dr. Hahn replied that the data supplied were difficult to reconcile with the results of his own experiments, made directly with a preparation of radiothorium, which seemed to show a half-value period of about two years for this substance. He made, however, at the same time the interesting suggestion that the lack of agreement could be explained if a rayless product having a slow rate of change intervened between thorium and radiothorium.

The rate of disintegration of radiothorium has since been determined by Blanc,[25] who finds that the half-value period is 737 days, and the existence of a product intermediate between thorium and radiothorium has recently been demonstrated by Hahn[26] in a very convincing manner.

Hahn made a careful examination of a series of samples of the intermediate chemical products obtained in the technical preparation of pure thorium nitrate from monazite by the firm of Dr. O. Knöfler and Co., in Berlin. He found only insignificant differences in the

---

[25] Rend. della R. Accad. d. Lincei, xvi, 291, 1907.
[26] Berichte d. chem. Ges., xl, 1462, 1907.

specific activity of the thorium series in these samples and no differences indicating the separation of any considerable proportion of the radiothorium.[27]

On testing samples of the pure thorium nitrate which had been prepared by the same firm some years before, he found that these showed a conspicuously lower activity than the freshly prepared salts, and this decrease in activity appeared to continue for about three years, after which the activity remained fairly constant for some time. He also states that he has been able to obtain preparations which are free from thorium, but which show with the passage of time a marked increase in their activity and in their power to produce thorium emanation. These preparations must contain the intermediate product between thorium and radiothorium. He therefore reaches the conclusion that it is this intermediate product, for which he suggests the name "mesothorium," and not the radiothorium, which is separated from the thorium in the technical process of preparing pure thorium nitrate.

The films of thorium oxide which were originally used for the determination of the activities given in my earlier paper, and also a number of others which had been prepared and measured at the same time or shortly afterward, have been carefully preserved. The activity of these films has been recently re-measured in the larger of the two electroscopes described in the earlier paper and it has been possible to compare their present activities with their activities at the time of the first measurement. It was found that the specific activities of all the oxides has decreased by an amount equal to from 15 to 30 per cent. of their former values.

The results obtained are shown in the following table. In column I the decrease in the activity of each oxide is given in terms of its activity when first measured. In column II is given the approximate time which elapsed between the two measurements. In column III is given the decrease to be expected in the activity of each oxide if the activity had been falling at a rate corresponding to half-value in 737 days (the rate of decay of radiothorium). Oxides numbered 1 to 8 are the oxides given corresponding numbers in my earlier paper. Oxide No. 9 was prepared from thorianite by a chemical process differing only slightly from that used in preparing oxide No. 5. Oxides No. 10 and 11 were separated from certain technical products supplied by the Welsbach Company and obtained

[27] During the past year I have made an examination of similar chemical products kindly supplied by the Welsbach Company. The results obtained were similar to those obtained by Hahn, and no marked differences were noted in the thorium specific activity of the freshly prepared materials.

by them in the process of preparing pure thorium nitrate from monazite.

TABLE

Number	Source of oxide	I Decrease %	II Time days	III Decrease calc.
1	Mantle dust	18	500	38%
2	Welsbach nitrate	32	489	38
3	" "	15	347	28
4	" "	22	365	30
5	Thorianite	30	489	38
6	"	26	428	34
7	Monazite	26	428	34
8	Miner's oxide	30	365	30
9	Thorianite	30	408	33
10	Welsbach residue No. 3	23	331	27
11	Welsbach residue No. 4	19	331	27

It will be noted that none of the oxides has lost its activity at a rate greater than that corresponding to a fall to half-value in 737 days, while most of the oxides have lost their activity at a lower rate. The behavior of the oxides indicates that at the time they were first measured they each contained much less mesothorium than the quantity in equilibrium with the thorium present. Oxides numbered 2, 5, 8, 9 and 10 in particular must have contained but a small proportion of their equilibrium amounts of mesothorium since the rate of fall of their activity has so closely approached that of radiothorium itself.

The results which have been obtained clearly indicate that the low specific activity of the thorium series in the salts prepared by the Welsbach Company can not be attributed to any special peculiarity in the chemical methods employed in the Welsbach works. It also appears that the chemical separation of thorium from mesothorium can be effected without difficulty by a variety of reactions.

A portion of the Welsbach nitrate from which the three oxides Nos. 2, 3 and 4 had been indirectly prepared has been preserved in its original crystalline form. A small amount of this nitrate was converted directly into oxide by ignition, and the specific activity of this oxide was determined a few days later. This was done shortly after the second series of measurements of the older oxides had been carried out. It was found that the specific activity of the thorium series in the original nitrate is now practically the same as it was at the time when the first series of measurements was made. The observation of this fact at once suggested the possibility that even the precipitation of thorium as hydroxide from a solution of the nitrate is effective in separating thorium from mesothorium, since this was the only treatment to which the thorium in oxide No. 2 had been subjected.

Over two years ago I had made some experiments with the object of obtaining a more definite knowledge of the chemical behavior of thorium-X. At that time a considerable number of thorium-X residues had been prepared by the well-known method of precipitating the thorium as hydroxide from a solution of the nitrate, filtering off and washing the hydroxide, evaporating the filtrate to dryness, and gently igniting the residue to remove the ammonium salts present. The amount of thorium nitrate used in some of these experiments was as much as a hundred grams and the volume of the filtrate was often more than two liters. In these experiments it was always found that the residue, after the removal of ammonium salts, contained very appreciable amounts of thorium. This thorium was finally removed from the residue by a second treatment with ammonia in a solution of small volume. A number of such thorium-free residues had been prepared and measurements of their activity had shown a steady fall for a period of about 30 days. After 30 days the residues still retained a definitely measurable activity, which was comparatively small, however, and amounted to a leak of only a few divisions per minute in the electroscope. This residual activity, which was observed further in some instances for a period of about one week, appeared to be fairly constant during that period, after which its progress was not further followed. It was attributed at the time to the presence of a little radium in the original nitrate.

Several of these old thorium-X residues have been preserved, and as soon as the possibility that they might have originally contained mesothorium suggested itself they were introduced into the electroscope and their activities noted. The activity of each of them was found to have risen enormously, until it has now reached a value which is many times greater than the minimum value to which it had originally fallen. These thorium-free residues, weighing together not more than a few milligrams, have now an activity equal to that of several grams of thorium oxide of normal activity and spontaneously evolve an emanation which is identical with that of thorium, falling to half-value in 54 seconds and producing the characteristic thorium-active deposit. These residues therefore contain radiothorium (and its products) which has been formed by the disintegration of the mesothorium originally present. If the half-value period of mesothorium is about 7 years, as suggested by Hahn, then these residues must still contain mesothorium and their activity will continue to increase for some time longer.

It is obvious, therefore, that the chemical process first described by Rutherford and Soddy for preparing thorium-X from thorium, consisting in the precipitation of the thorium as hydroxide from a solution of the nitrate, can be employed also for the separation of mesothorium from thorium. It has the advantage that the mesothorium is obtained in a relatively concentrated form, but it can be applied with advantage only where thorium nitrate of some age is available, since fresh thorium nitrate will contain little if any mesothorium.

The fact that the two oxides numbered 3 and 4, which were obtained through the intermediate formation of the normal and "basic" sulphates, respectively, also show a decrease in their activities is not in itself very suggestive, since in the preparation of the sulphates the nitrate was first converted into hydroxide. Also in the cases of the other oxides, No. 5 to No. 11 inclusive, the chemical operations through which the thorium passed from the first decomposition of the mineral to the final separation of the pure thorium salt were too complicated to make it possible to determine at what particular step the mesothorium was removed. From various data it appears probable, however, that the precipitation of thorium by sodium thiosulphate is effective in separating thorium from mesothorium.

A further point which appears to be worthy of notice in passing is the similarity in chemical behavior shown by thorium and radiothorium on the one hand, and by thorium-X and mesothorium on the other. We have no good evidence as yet of the chemical separation of thorium and radiothorium. These two elements appear to remain together most persistently through elaborate chemical operations which result in the separation of the thorium from every other known element. In contrast to this is the facility with which thorium-X and mesothorium can be separated from thorium and radiothorium. The chemical similarity of mesothorium and thorium-X is further indicated by what follows. Nearly two years ago an attempt was made to separate radiothorium from thorium by precipitating barium sulphate in a dilute solution of a thorium salt. This experiment was performed because it was thought that the entrainment of radiothorium by barium sulphate might explain the presence of radiothorium in the radium-barium sulphate residue for thorianite where Hahn had first obtained it. The precipitated barium sulphate was highly active when first prepared, but its activity fell regularly at a rate corresponding to half-value in 4 days, until at the end of about 40 days it had reached a negligible value. At the start it therefore contained thorium-X but no appreciable amount of radiothorium. This precipitate of barium sulphate has recently been examined and is now quite active. It gives off thorium emanation but contains no thorium. Its present activity is therefore due to radiothorium formed from the mesothorium which was precipitated with it. From this it seems probable that the radiothorium which Hahn separated from the residue of insoluble sulphates, obtained some time before in working up a considerable quantity of thorianite, had not been present in the residues when first prepared, but had been formed in them later through the disintegration of mesothorium. This supposition is further strengthened by Hahn's statement[28] that his radiothorium probably contained some mesothorium.

It appears quite likely, therefore, that the entraining action of barium sulphate on mesothorium was directly responsible for the presence of radiothorium in the thorianite residues.

Summary.

Measurements of the changes which have taken place in the activity of certain thorium preparations have given results which are strongly in support of Hahn's assertion that an intermediate product having a slow rate of change and not emitting α-rays exists in the thorium series between thorium and radiothorium.

Certain methods for the separation of this intermediate product from thorium have been described.

Sloane Laboratory, Yale University,
New Haven, Conn., June 16, 1907.

[Herbert N. McCoy (1870–1945) of the University of Chicago had a talent, which is not altogether common, for precision in measurement. He had made careful determinations of the total radioactivity of various minerals and salts, and had found in each case an activity strictly proportional to the content of uranium or thorium.[29] On the theory of radioactive equilibrium, that meant that every radioactive substance in each specimen was a transformation product of one or the other of those primary materials. To measure the radioactivity, he had used alpha-particle ionization, and since alpha particles are easily absorbed, he had taken pains in the preparation of his samples to keep the absorption uniform and at a minimum. The paper reprinted here is one of a pair which he published in the fall of 1907 in which the earlier work was

---

[28] Loc. cit.

[29] H. N. McCoy, "Ueber das Entstehen des Radiums," *Ber. dtsch chem. Ges.*, 1904, *37*: 2641–2656; "Radioactivity as an Atomic Property," *J. Amer. chem. Soc.*, 1905, *27*: 391–403; "The Relation between the Radio-activity and the Composition of Uranium Compounds," *Phil. Mag.* [6], 1906, *11*: 176–186.

H. N. McCoy and W. H. Ross, "The Relation Between the Radio-Activity and the Composition of Thorium Compounds," *Amer. J. Sci.* [4], 1906, *21*: 433–443.

repeated for the sake of improved geometry in the ionization measurements.[30]

In the case of thorium, there had been complications which Hahn's discovery of mesothorium was just beginning to explain. The ratio of alpha-particle activity to thorium content was constant for all minerals and for thorium oxide freshly prepared from a mineral. It was lower for thorium oxide prepared from commercial thorium nitrate. Both thorium and radiothorium contributed to the alpha-particle activity, and since their chemistry was similar, every extraction of thorium extracted radiothorium as well, while leaving the mesothorium behind. A newly prepared thorium sample would still contain radiothorium in large quantity, but as time went on two processes would alter its composition. The radiothorium, transmuting into thorium X and the short-lived products which followed, would gradually disappear. At the same time, the transmutation of the thorium would begin to replace the lost mesothorium and the mesothorium in its turn would transmute to fresh radiothorium. Thus any refined preparation of thorium should show an alpha-particle activity which was high at first, decreased to a minimum over a period of a few years, and slowly rose toward a final maximum.

Since they understood all this, McCoy and Ross did not limit their investigations simply to improved measurements of radioactivity. They hunted for a method of separating radiothorium from thorium so that each could be studied apart from the other. They measured the activity of thorium preparations of known age, looking for the minimum value; with this known, they could calculate the half-life of mesothorium, using Rutherford's mathematical theory. Here they succeeded easily, obtaining a half-life of 5.5 years. In their other project they succeeded in fact, although not in a manner evident at that time. They found no technique which was in the least effective, and after a good deal of vigorous effort they could conclude only that "The direct separation of radiothorium from thorium by chemical processes is remarkably difficult if not impossible."—A. R.]

[30] The citation for the other is: H. N. McCoy and W. H. Ross, "The Specific Radioactivity of Uranium," *J. Amer. chem. Soc.*, 1907, *29*: 1698–1709.

15

Herbert N. McCoy and W. H. Ross

The Specific Radioactivity of Thorium and the Variation of the Activity with Chemical Treatment and with Time

[From *The Journal of the American Chemical Society*, 1907, *29*: 1709–1718.]

In our first paper[31] on the radioactivity of thorium compounds we showed that, in the case of minerals, when the activity due to uranium was deducted, the remaining activity was strictly proportional to the percentage of thorium present. The following table, taken from the earlier paper, summarizes the results obtained:

TABLE 1.

No.	Name	% Th	% U	k_m	k_U	k_{Th}	$\frac{k_{Th}}{P_{Th}}$
1	Orangite	43.1	7.76	649	255	394	914
2	Thorite	46.6	6.26	664	205	459	985
3	Monazite	5.27	0.33	60.2	10.9	49.3	935
4	"	15.18	0.46	164	15	149	982
5	"	2.72	0.12	29.8	4.0	25.8	950
						Mean =	953

The symbol k_m represents the specific activity of the mineral, in terms of the activity of 1 sq. cm. of a thick film of U_3O_8 as unity; k_U is the activity due to uranium; $k_{Th} = k_m - k_U$. P_{Th} is the weight of thorium per gram of mineral.

Results similar to these were published by Boltwood[32] in the same number of the Am. J. Sci. as that in which our paper appeared. Boltwood's results were expressed in arbitrary units, but showed just as clearly as do those of Table 1, the constant specific activity of thorium in minerals. A third simultaneous paper, in the same journal by Dadourian[33] contained similar conclusions reached as a result of a

[31] Am. J. Sci., **21**, 433 (1906).
[32] Ibid., **21**, 409 (1906).
[33] Ibid., **21**, 427 (1906).

different method of investigation; which consisted in the measurement of the intensity of the excited activity obtained from the thorium emanation given off by solutions of minerals. These results were also expressed in arbitrary units.

In the first paper we stated:

"All of the thorium activity measurements, as well as those of uranium compounds and minerals, were made with a distance of 3.5 cm. between the active films and the charged electrode. While this thickness of air is sufficient to absorb practically all of the alpha rays of uranium, it is scarcely great enough to absorb completely the more penetrating alpha rays of some of the radium and thorium products.[34] With a greater distance than 3.5 cm. between the films and electrode a somewhat greater activity will be found for uranium and thorium minerals. The general relationship between the radioactivity and the composition of such minerals having now been fully established, we are starting a new series of measurements on minerals with the object of determining their activities under conditions such that the maximum ionizations due to the alpha rays can take place in the space between the film and the electrode." The recent work of Hahn[35] on the ranges of the products of thorium, also emphasizes the necessity of a larger ionization chamber than we first used.

In the new series of measurements, we worked in the manner described in the preceding paper on *The Specific Radioactivity of Uranium*, using the new electroscope, and films of the mineral deposited by sedimentation on flat, rimless plates. Since the earlier work had shown that the specific activity of thorium in its minerals is constant it was considered unnecessary to repeat the activity measurements of more than three of the minerals in order to determine the exact value of this constant. The new results are given in Table 2. The new value found for the specific activity of uranium in minerals, 3616, was used in calculating the activity due to that element in each thorium mineral.

TABLE 2.

Name	$\% Th$	$\% U$	k_m	k_U	k_{Th}	$\frac{k_{Th}}{P_{Th}}$
Orangite	43.1	7.76	707	281	426	988
Thorite	46.6	6.26	699	226	473	1015
Monazite	15.18	0.46	172	16.6	155.4	1025
				Mean	=	1009

[34] Rutherford, Radioactivity, p. 168, 1905; Bragg and Kleeman, Phil. Mag., **10**, 318 and 600 (1905).

[35] Phil. Mag., **11**, 794; 12, 82 (1906).

The mean value of k_{Th}/P_{Th}, 1009, is the specific activity of thorium containing the equilibrium amounts of its products.

We also previously found that the specific activity of thorium in pure ThO_2 obtained in the analyses of minerals by Neish's process,[36] was constant and equal to the specific activity of thorium in minerals. But the activity[37] of pure ThO_2 prepared from commercial samples of pure nitrate was always much less than that obtained from minerals by Neish's process.[38] The latter sort contained the maximum or equilibrium amount of radiothorium, while the former evidently did not. It seemed, therefore, as if the technical method of preparation of thorium nitrate was more effective in separating radiothorium from thorium than was Neish's analytical method. We therefore attempted, by various chemical processes, to remove from thorium the whole of the radiothorium, with the object in view of obtaining thorium of minimum activity. As previously announced, we were able by the application of certain processes, many times repeated, to reduce the permanent activity of thorium considerably below that of the least active commercial sample used. But, in spite of the apparently promising results of the preliminary experiments, we did not succeed in separating radiothorium completely from thorium. In fact, it now appears doubtful whether it is possible by chemical treatment to separate any radiothorium from thorium. In a recent paper,[39] Hahn has pointed out that the activity of thorium preparations decreased spontaneously to an appreciable extent in the course of a few years; Boltwood[40] has also observed the same thing. We are able to confirm these observations by new measurements of the activity of preparations made and first measured about one and one-half years ago. Hahn also found thorium preparations nine to 12 years old were more active than those three years old. These facts are explained by Hahn by the supposition that there is an inactive product, which he has called mesothorium, between thorium and radiothorium and that this intermediate product is removed in the process of preparation of pure compounds from minerals; but that the radiothorium remains with the thorium. Hahn first estimated the period of mesothorium to be about seven years; in his second paper[41], he states that his estimate is apparently somewhat too high, but in

---

[36] This Journal, **26**, 780 (1904).

[37] The activities were measured four or five weeks after preparation when the samples had regained the amounts of ThX, etc., corresponding to the amounts of radiothorium present in each case.

[38] Boltwood and Dadourian obtained similar results.

[39] Ber., **40**, 1462 (1907).

[40] Am, J. Sci., **24**, 93 (1907).

[41] Ber., **40**, 3304 (1907).

neither paper does he give the numerical data upon which the estimate is based. According to Blanc,[42] the period of radiothorium is 737 days. The low activity of commercial samples of thorium, as observed by Boltwood and by Dadourian, as well as by ourselves, is explained by Hahn as due to the decay of radiothorium with time, rather than to its removal by chemical processes. It is also probable that the decrease in activity of thorium preparations, which we observed after numerous chemical treatments, was the result solely of the decay of radiothorium, (aided by the chemical removal of mesothorium), during the time which had elapsed between the measurements of activity which preceded and those which followed the chemical treatments; in fact our experiments strongly indicate that radiothorium is entirely inseparable from thorium by chemical processes. Of course the indirect separation of radiothorium may doubtless be accomplished by the removal, from time to time, of the mesothorium; the radiothorium will then continue to decay until in the course of 15 or 20 years it has practically completely disappeared.

We found, as shown in our first paper on this subject, that thorium dioxide obtained in the analyses of minerals by Neish's method had the same activity as it had in the original mineral. Neish's process involves one precipitation with oxalic acid; one with potassium hydroxide; and two with nitrobenzoic acid. Boltwood found similar results with thorium dioxide obtained from minerals by analytical processes in which the thorium was separated either by means of repeated treatments with ammonium oxalate or repeated precipitations with sodium thiosulphate. It follows from these results that none of the chemical processes involved removes any appreciable fraction of the radiothorium.

We have made a number of additional experiments in which commercial samples ("A" and "B") of thorium nitrate[43] have been subjected to the various treatments described below and finally converted into thorium dioxide.

(1). Twenty grams of sample A were dissolved in 200 cc. of boiling water and a solution containing one gram of oxalic acid was added; this precipitated about one-eighth of the thorium. The oxalate so obtained was converted into oxide by ignition.

(2). Additional one gram portions of oxalic acid were added to the filtrate from (1), the precipitate being filtered out after each

---

[42] Physik. Z., **8**, 321 (1907).

[43] The samples were made by the firm of C. A. F. Kahlbaum; the oxide obtained by ignition was analyzed by Neish's method and found to be practically chemically pure in each case.

addition; the eighth precipitate was ignited to oxide and constituted sample 2.

(3). Ten grams of sample A were dissolved in 300 cc. of boiling water and precipitated with 12 g. of sodium thiosulphate in 100 cc. of water. The boiling was continued one or two minutes. The precipitate was filtered out and dissolved in dilute nitric acid, the excess of the latter was nearly neutralized with ammonia and the whole process repeated. After eight precipitations in this manner the material was converted into oxide.

(4). Five grams of sample A were dissolved in 200 cc. of water and precipitated, at boiling temperature, with five grams of potassium chromate in 50 cc. of water. The precipitate was filtered out and dissolved in dilute nitric acid; the solution was diluted and the thorium chromate again precipitated from the hot solution by the addition of ammonia. The solution was acid, in reaction, at the end of the precipitation. After eight precipitations in this way, the solution in nitric acid was treated with an excess of oxalic acid; the oxalate was washed and converted into oxide.

(5). Two grams of the sample A and 10 g. of barium chloride were dissolved in 200 cc. of water and a solution of 6.5 g. of ammonium oxalate was added; this was just sufficient to precipitate both the thorium and the barium as oxalates. Another portion of 6.5 g. of ammonium oxalate was now added to dissolve the thorium as double oxalate. After being heated to boiling, the residue of barium oxalate was filtered out and the filtrate was treated with 10 g. more of barium chloride. This process was repeated until 7 precipitations had been made. The barium oxalate was removed from the last precipitate by means of dilute hydrochloric acid: the thorium oxalate remaining was converted into oxide by Neish's method.

(6). Ten grams of sample A were dissolved in 400 cc. of water and precipitated with an excess of freshly distilled ammonia. The separated precipitate was dissolved in nitric acid and the process repeated 100 times; the last precipitate was converted into oxide. Analysis showed that this oxide contained 4.6 per cent. of impurity, due to the action of the alkaline solutions on the glass vessels used.

(7). Twenty grams of solid nitrate, sample B, were ground in a mortar with an excess of concentrated ammonia; the hydroxide so formed was filtered out, and ignited to oxide. The oxide so obtained is much more compact than that formed by the direct ignition of the nitrate.

(8). Fifteen grams of sample B were dissolved in 300 cc. of water and precipitated with hydrogen peroxide, at a temperature of about 70°. The separated precipitate was dissolved in nitric acid; the thorium was again precipitated from the dilute solution by the addition of 45 g. of ammonium acetate and 35 cc. of hydrogen peroxide. The whole process was repeated until 40 precipitations had been made; the last precipitate was purified by Neish's method and converted into oxide.

The activities of these preparations were measured in the old way (films in tins with rims and 4.5 cm. between films and electrode), but the results as given in Table 3 have been multiplied by the factor 1.12, which is the ratio of the activity as determined in the new way (flat films and 8.5 cm. ionization space) to that determined in the old.

TABLE 3—ACTIVITIES OF SAMPLES OF THORIUM DIOXIDE

	Chemical treatment	Date of precipitation. Oxides from sample A	Activity on May 8th, 1906	Activity on September 12th, 1907
1.	1 pptn. oxalic acid . .	Apr. 6, 1906	529	417
2.	8 pptn. oxalic acid . .	Apr. 6, 1906	529	423
3.	8 pptn. thiosulphate .	Apr. 4, 1906	529	414
4.	8 pptn. chromate	Apr. 9, 1906	547	421
5.	7 pptn. barium oxalate	June 11, 1906	. . .	420
6.	100 pptn. ammonia	Apr. 3, 1906	510[44]	402[44]
		Oxides from sample B	Feb. 24, 1906	Sept. 12, 1907
7.	1 pptn. ammonia . . .	Nov. 11, 1905	477	410
8.	40 pptn. hydrogen peroxide	June 28, 1906	. . .	385

As has been shown, one precipitation with oxalic acid, (which is included in Neish's analytical method), does not remove any radiothorium from thorium. Since the activities of samples (2) to (5) were practically the same as that of sample (1) at each time of measurement, as shown in Table 3, it follows that none of the processes used in making these samples effects any separation of radiothorium. The activity of sample (6) is a little low, but the difference, about 4 per cent., may be only experimental error; at most since there were 100 precipitations, one precipitation with ammonia removes but 0.04 per cent. of the radiothorium present. It would seem from a comparison of the activities of (7) and (8) that 40 precipitations with hydrogen peroxide remove a small portion of the radiothorium; yet the lower activity of (8) may be due to a different cause: the incomplete removal of mesothorium from (7) at the time of its preparation from the nitrate Nov. 11, 1905. This point will be considered further in subsequent

---

[44] The values given have been corrected for the 4.6% impurity contained in the sample.

paragraphs. The isolation of radiothorium from thorianite[45] and from pure thorium nitrate[46] seems to indicate the possibility of separating radiothorium by chemical processes, but these results may have been accomplished through the separation of mesothorium, which in time changed spontaneously into radiothorium.

The specific activities given in Table 3 refer to thoria containing the amounts of ThX and subsequent products, corresponding to the amounts of radiothorium in the samples, since the activity measurements were made about one month after the last chemical treatment. As most of the chemical processes, used in the attempts to remove radiothorium, readily remove ThX, the activity of any sample immediately after its preparation was much less than one month later. Rutherford and Soddy,[47] who were the first to observe this fact in the case of thorium precipitated with ammonia, found that the final activity was about 4 times as great as the initial. We found the ratio of final to initial activity to be approximately 2.5 for the various samples of thoria prepared from sample A of nitrate.

The corresponding ratio for thoria separated from minerals in analyses by Neish's method was about 3.2. The greater ratio in the case of thoria rich in radiothorium was observed by us at the time of the measurements made in May, 1906, Table 3, and was considered, at that time, to prove conclusively that thorium itself is active. And, although the available data were not exact, we estimated that the specific activity of thorium dioxide alone was probably between 100 and 130. Hahn, in his recent paper already referred to, has observed the greater ratio of maximum to minimum activity for samples rich in radiothorium and has pointed out that this shows finally and conclusively that thorium itself is active, that is, that thorium alone emits α-rays. We do not wish to make any claim of priority on this point, as this is the first publication of our observations; but we do wish to emphasize the fact that Hahn's conclusion is correct.

As was just stated, the data from which we first estimated the activity of thorium alone were not exact; this was due to the fact that the numerous precipitations with ammonia required for the removal of ThX and its products[48] introduced silica and other impurities from the glass. An improved method for the separation of ThX and all subsequent products from thorium has been worked out by Mr. G. C. Ashman and one of us.[49] It was found that precipitation of thorium with meta-nitrobenzoic acid leaves ThX and ThA in the solution;

[45] Ramsay, J. chim. phys., **3,** 617 (1905); Hahn, Ber., **38,** 3372 (1905).
[46] Blanc, Physik. Z., **7,** 620 (1906).
[47] Phil. Mag., Sept., 1902.
[48] Rutherford and Soddy, Loc. cit.
[49] The results will be published soon.—H. N. McCoy.

four precipitations at intervals of two hours give finally chemically pure thorium dioxide, which is entirely free from ThX and its subsequent products. From the observations of the minimum and maximum activities of the two samples of thoria, one containing but little radiothorium, the other freshly prepared from a mineral, the specific activity of thorium dioxide, free from radiothorium and all other active products, was found to be 105. The activities of the equilibrium amounts of radiothorium and of ThX and subsequent products for one gram of thorium dioxide were 182 and 655 respectively. These values refer to flat films and an ionization space of 8.5 cm. between the film and the electrode, the measurements having been made in the electroscope described in the preceding paper.

If the chemical processes involved in the preparation, from a mineral, of any sample of a pure thorium compound remove all or part of the mesothorium, the activity of the sample will fall to a minimum, in the course of time, and then increase again, as Hahn has stated. It follows also that this minimum activity will be *perfectly definite*, provided the whole of the mesothorium is removed and that the purification processes are all carried out at one time, say within a period of one month. The ratio of this minimum activity to the activity of thorium containing the equilibrium amounts of all its products is a fraction, which is a function of (1) the period of mesothorium; (2) the period of radiothorium; (3) the activity of thorium alone; (4) the activity of thorium and its subsequent products. Conversely, if the value of this minimum be known, the period of mesothorium may be calculated, since all of the other data are now available.

Rutherford[50] has developed the general equation for the variation with time of the number of particles of one radioactive body, produced by the disintegration of another. This equation applies to the formation of radiothorium from mesothorium. Let λ_1 and λ_2 represent the disintegration constants and n_0' and n_0'' the equilibrium numbers of particles per gram of thorium dioxide, of mesothorium and radiothorium respectively. The number of particles, n'', of radiothorium formed in time, t, from n'_0 particles of mesothorium, which contained no radiothorium at time zero, is given by equation (1)

$$n'' = \frac{n_0'\lambda_1}{\lambda_2 - \lambda_1}\left(e^{-\lambda_1 t} - e^{-\lambda_2 t}\right) \tag{1}$$

But

$$n_0'\lambda_1 = n_0''\lambda_2$$

Therefore

$$n'' = \frac{n_0''\lambda_2}{\lambda_2 - \lambda_1}\left(e^{-\lambda_1 t} - e^{-\lambda_2 t}\right) \tag{2}$$

[50] Radioactivity, 331, (1905).

If, from a quantity of thoria which contained the equilibrium amounts of all its products, the mesothorium only were removed completely at time zero, then $(n_0'' - n'')/n_0''$ would represent the fraction of the equilibrium number of particles of radiothorium left in the thoria at the time, t. The fraction, $(n_0'' - n'')n_0''$, decreases with time and reaches a minimum, which may be called m, at a time which depends only on the periods of the two substances concerned. It we take the activity of one gram of thorium dioxide alone as 105 and the activity of the equilibrium amounts of its products as 837 the minimum specific activity of the thoria, corresponding to a minimum content of radiothorium, is equal to $837m + 105$.

We have found that the present activity of a sample of thoria, made by strongly igniting sample A of thorium nitrate, is 476. If we take this activity as the minimum, we may write $837m + 105 = 476$; from which $m = 0.443 = (n_0'' - n'')n_0''$. From this last relationship and equation (2), the value of λ_1 may be found; and from λ_1 the period of mesothorium. The value thus found for the last mentioned constant was 5.4 years; it may actually be greater, however, since the minimum activity may be less than 476. If the period is 5.4 years, the minimum activity will be reached in 4½ years from the date of preparation. Since the specific activity of this sample was 579 on Dec. 8, 1905, it was then probably about 2.0 years old. If so, it is now, Sept., 1907, about 3.8 years old; and although the minimum activity will not be reached for 9 or 10 months (assuming the period of mesothorium to be 5.4 years) the total activity will diminish less than one per cent. in this time. From these considerations it seems probable that the period of mesothorium is not far from 5.5 years. If so, the minimum activity will be reached in 4.6 years. This conclusion is in good agreement with the changes of activity of samples which have been under observation for about 1½ years.

On April 4th to 9th, 1906, samples (1) to (4), Table 3, were subjected to chemical treatment which probably removed all of the mesothorium present at that time. The mean activity of these samples was 534 on May 8, 1906, and 418 on Sept. 12, 1907. In the interval, 492 days, the radiothorium present on May 8th, 1906, would decay to $e^{-492\lambda_2} = 0.630$; the products of radiothorium would disappear in the same proportion. Now the specific activity of thorium dioxide itself is 105. If no radiothorium had been formed from the new mesothorium produced from thorium during the interval of about 525 days between the date of preparation and that of final measurement, Sept. 12, 1907, the activity of the samples would have been $0.630(534 - 105) + 105 = 375$ on the latter date. But some radiothorium must have been produced from the mesothorium formed from thorium during this interval and it is easy to show that the activity of the radiothorium so formed would be, for $t = 525$,

$$837\left\{1 - \left[\frac{\lambda_2}{\lambda_2 - \lambda_1}\left(e^{-\lambda_1 t} - e^{-\lambda_2 t}\right) - e^{-\lambda_2 t}\right]\right\} = 30.$$

The calculated specific activity on Sept 12, 1907, of samples 1 to 4 is, therefore, 375 + 30 = 405. The observed activity was 418.

Table 4 gives a summary of the results, of a similar sort, obtained with other samples already described as well as with two additional samples, 9 and 10.

Sample 9 was thorium dioxide prepared by Neish's method from a sample of "partially purified oxalates," made by the Welsbach Light Company.[51]

Sample 10 was thorium dioxide obtained in the analyses of orangite and thorite[52].

TABLE 4

Samples	Date of Preparation	Date and Activity		Date and Activity		Calculated Activity
1-2-3-4	Apr. 4-9, 1906	May 8, 1906;	534	Sept. 12, 1907;	418	405
6	" 3, "	" " "	510	" " "	402	390
7	Nov. 11, 1905	Feb. 24, "	477	" " "	410	370
9	May 8, 1906	July 6, "	854	" 23, "	610	625
10	Dec. 2-6, 1905	Feb. 9, "	947	" 28, "	621	630

With the exception of the results for sample (7), the values calculated and observed are in fairly good agreement. The higher observed activity of (7) is very probably due to incomplete removal of mesothorium at the time of the preparation of the oxide from the nitrate, Nov. 11, 1905; since the chemical treatment consisted only in grinding the solid nitrate with concentrated ammonia, filtering and igniting. The difference in activity of (7) and (8), Sept., 1907, (Table 3) may also be due to the incomplete separation of mesothorium from (7), rather than to the removal from (8) of part of the radiothorium by the chemical treatment employed.

On the other hand, it is very probable that the chemical processes carried out in the preparation of all the other samples, removed practically all of the mesothorium contained in them. Thus the activity of a sample of thorium dioxide, made by ignition alone of

[51] We wish to express our thanks to Mr. H. S. Miner, chemist for the Welsbach Company, who has supplied us with this as well as a number of other samples of thorium products.

[52] See Table XIV, Am. J. Sci., **21**, 443 (1906).

sample A of nitrate, had an activity of 476 in Sept., 1907, while samples 1 to 4, made from the same sample of nitrate, had a mean activity of 418.

Summary

1. The specific activity of thorium in minerals was found to be 1009 (the unit being the activity of one sq. cm. of a thick film of U_2O_8) when measured under such conditions that all of the α-rays reach their maximum ranges in air.

2. The direct separation of radiothorium, from thorium by chemical processes is remarkably difficult, if not impossible. This question is being studied further in this laboratory.

3. The diminutions of activity of thorium, which we previously observed after certain chemical treatments, are completely explained by Hahn's hypothesis of an intermediate product, mesothorium, between thorium and radiothorium; which product is easily removed by the chemical treatments; in consequence of which the radiothorium decays with time and thus causes the observed changes of activity.

4. If all of the mesothorium be removed, at one time, in the process of preparation from a mineral of pure thorium dioxide, the activity of the latter will fall to a definite minimum. From the activity of a sample of thoria, supposed to have reached this minimum, the period of mesothorium was calculated to be 5.5 years.

5. The quantitative changes of activity with time, of a number of samples of thoria agreed well with the values calculated on the assumption of a period of 5.5 years for mesothorium.

Oct. 12, 1907.

[The next paper, by Otto Hahn, established another link in the genetic chain of thorium which was to be of some importance in the final development of the transformation theory. He had found a beta-ray activity with the long-lived mesothorium he had discovered. Now it appeared that the beta radiation belonged to a quite different, short-lived substance which was chemically separable from the mesothorium.

With this proliferation of thorium descendants, Hahn proposed a systematic nomenclature modeled on the Curie-Rutherford scheme of radium A, radium B, radium C. The first transformation product of thorium would be *thorium 1* (his original mesothorium), the second, *thorium 2* (the new beta emitter), the third *thorium 3* (his original radiothorium), and so on. This system was described in a separate note[53] and

[53] O. Hahn, "Zur Nomenklatur der Thoriumverbindungen," *Phys. Z.*, 1908, *9*: 245. [This is the title printed at the head of the note. It seems to have originated in an editorial misunderstanding and is corrected on p. 320 of the volume to "Zur Nomenklatur der Thoriumzerfallsprodukte."—A. R.]

summarized in a table at the end of the paper reprinted here. It was never generally adopted; so long as genetic names remained in use most writers preferred the *mesothorium 1*, *mesothorium 2*, and *radiothorium* which appear in parentheses in the table.

As regards Hahn's chemistry, the modern reader will notice that thorium 2 or mesothorium 2 is Ac^{228}. The method of precipitating with ammonia was precisely the one which Rutherford and Soddy and later Boltwood had found effective for separating the thorium isotopes (thorium and radiothorium) from the radium isotopes (thorium X and mesothorium 1.) Hence any radiothorium in the original preparation would inevitably accompany the mesothorium 2, and the activity of the radiothorium and its succession products must be taken into account, as Hahn indicates in some detail.—A. R.]

16

Otto Hahn

A Short-Lived Intermediate Product between Mesothorium and Radiothorium

[Translation of "Ein kurzlebiges Zwischenprodukt zwischen Mesothor und Radiothor," *Physikalische Zeitschrift*, 1908, *9*: 246–248 (15 April).]

In a paper on a new intermediate product in thorium[54] I stated that the transformation of thorium into radiothorium did not take place directly, but that we must assume an intermediate product with a disintegration period longer than that of radiothorium. This substance was separated from monazite sand residues and also from old thorium salts; over a period of months it showed a constant rise in activity, and this rise was due to the formation of radiothorium, which can be detected unmistakably by the characteristic emanation. I called the substance mesothorium.

In a later work, "On the Rays of the Thorium Products,"[55] I made the statement that the new substance gave out beta radiation whereas thorium itself emitted alpha particles.

In the course of further investigations, it has now been shown that mesothorium is not a homogeneous substance but is composed of two constituents which stand in a genetic relationship; indeed the long-lived mesothorium proper disintegrates without perceptible radiation

[54] Ber. d. D. Chem. Ges. **40**, 1462.
[55] Ber. d. D. Chem. Ges. **40**, 3304.

into a new product which gives out the beta rays mentioned above[56] and possesses a period of disintegration of 6.20 hours.

In a mixture of the various disintegration products of thorium the new product can easily be overlooked; in addition its purification—that is to say its separation from all other disintegration products of thorium—is hardly successful. It can be achieved, however, if one starts with mesothorium which is as pure as possible in the radioactive sense. In such a preparation the division of the original mesothorium into two homogenous components succeeds without difficulty.

In the Note which preceded this publication I have stated that I propose for mesothorium proper, as the first disintegration product of thorium, the name of thorium 1, and for the new substance, thorium 2.

The separation of thorium 2 from thorium 1 occurs conveniently with ammonia. Thorium 2 separates while thorium 1 remains in solution. It is necessary of course to add to the solution being treated a substance which is precipitated by ammonia in chemically recognizable amounts, since the radioactive substance to be separated by ammonia occurs in an unweighable and invisible amount.

For various reasons I have decided on zirconium as thoroughly suitable to serve as a deposition medium in the precipitation with ammonia. A few drops of a zirconium chloride solution were added to the precipitating solution; on the addition of pure ammonia the thorium 2 precipitates with the ammonia. It can then be separated from thorium 1 by filtration and studied with respect to its radioactive properties.

In this, however, the following is to be noticed. Mesothorium forms radiothorium or thorium 3 (see the preceding Note) with its characteristic period; almost simultaneously, thorium X and the other disintegration products come into being. If one precipitates by the method given above a mesothorium preparation which has been left to itself for a short time, the newly formed radiothorium and the active deposit are precipitated with the ammonia. While radiothorium itself is of no significance for the disintegration curve of the beta rays from thorium 2, this is not the case for the active deposit which is separated with the radiothorium and also indirectly formed by it, since it shows the very penetrating beta rays of thorium C. This beta radiation will partly cover the weaker rays from thorium 2 and thus affect the curve.

For this reason, the earlier values found for the decrease were too

[56] A joint investigation with Dr. Lise Meitner comparing the beta rays of mesothorium with those of other radio-products is in progress and will be published shortly.

high and did not follow a simple exponential law. More recent experiments with mesothorium which had been carefully and repeatedly purified have given very consistent values and a uniform curve of decrease for thorium 2.

Figure 1 gives the characteristic curves for the decay of a pure preparation of thorium 2 (curve A) and the formation of thorium 2 from pure thorium 1 (curve B). The circles give the experimental values, which are scaled to 100 for the initial value of curve A and the final value of curve B.

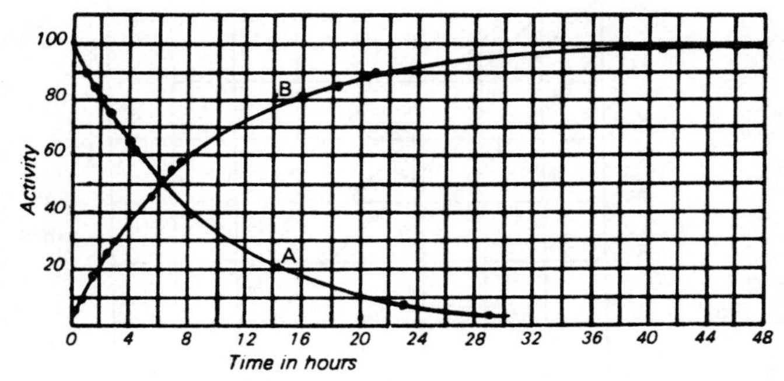

Fig. 1

Neither for the decay curve nor for the recovery curve has any residual activity or initial activity been deducted; the values found for the beta rays have been plotted directly. This indicates on the one side that the mesothorium used possessed a high degree of radioactive purity, and in addition that the separation of the two components proceeds very smoothly under the proper conditions.

I should not omit to mention that in the recording of the disintegration curves of thorium 2 small fluctuations were often observed at the beginning whose cause I do not know. Nevertheless in most cases the decrease was exponential even from the beginning so that I attribute no significance to these unimportant fluctuations.

Immediately after the separation of thorium 2 from thorium 1, the residual beta activity left with the latter amounted in the cases cited to no more than 2 per cent and probably less.

In Figure 2 a number of decay curves are plotted which were taken with different preparations of thorium 2. The ordinates give the logarithm of the activity, the abscissas the time in hours.

It can be seen that the values found lie in good agreement along straight lines, corresponding to the disintegration of a homogeneous radioactive substance. Only the later values of the fourth curve show an unmistakable departure from the line, which arises in this case from a small quantity of the active deposit of thorium. That is to say, the mesothorium used for the separation of thorium 2 had already stood for a short time and formed a certain quantity of radiothorium + disintegration products, out of which the active deposit was separated with the thorium 2.

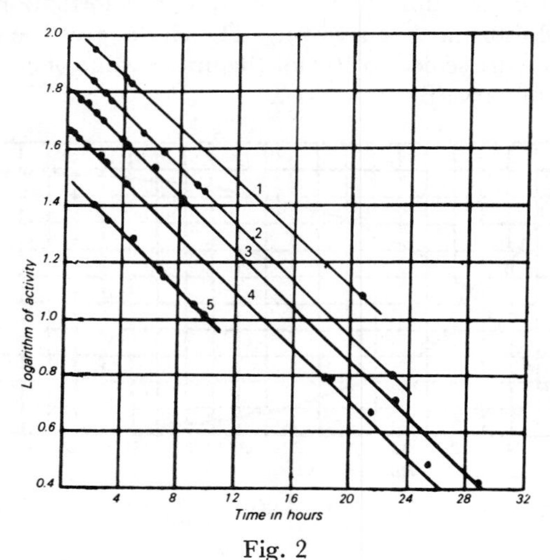

Fig. 2

Still older preparations of thorium which were originally pure gave curves for the separated thorium 2 which departed still farther from straight lines. The curves then represent mixtures of thorium 2 with thorium A + B + C. I have dispensed with the reproduction of such curves since the relationship is completely clear.

In the following brief table the disintegration constants and the corresponding values of the disintegration period are given, as obtained from a series of determinations: Nos. 1–5 correspond to values from the disintegration curves of thorium 2 given in Figure 2; Nos. 6 and 7 correspond to numbers obtained from recovery curves for thorium 1 freed of thorium 2 (one of which is designated as Figure 1B.)

Disintegration and recovery curves give good agreement, and the mean value of 6.20 hours should be correct within narrow limits.

Within 2–3 days after its preparation the thorium 2 has disappeared and the thorium 1 is as strong as at first. Taken strictly, to be sure, the relations are more complicated. Thorium 2 as the second component of the original mesothorium must represent the mother substance for radiothorium or thorium 3. Thus the beta activity of a

	λ	Disintegration period
1	0.1073	6.45 hours
2	0.1133	6.12 "
3	0.1115	6.22 "
4	0.1118	6.20 "
5	0.1123	6.17 "
6	0.1108	6.25 "
7	0.1118	6.20 "
Mean excluding the 1st value	0.1118	6.20 hours

thorium 2 preparation must not disappear completely; a trace should remain and reach a maximum in the course of a month.

Nevertheless it is easily calculated that in relation to the original activity of thorium 2, the increasing residual activity arising from newly formed thorium C gives only a negligible contribution which is impossible to detect under ordinary conditions.

The alpha particles behave somewhat differently. Thorium 2 gives off no alpha particles and disintegrates into radiothorium, which emits alpha particles and for its part produces in a short time four other alpha-ray products. Given absolutely pure thorium 2, it is to be expected that at first almost no effect will be observed in an alpha-ray electroscope, and that gradually an activity will become noticeable. It has already been mentioned that some of the radiothorium and thorium A + B + C present will be separated with the thorium 2. If the content of thorium X (or radiothorium X) in mesothorium amounts for example to only 1 per cent, the alpha activity which arises from that one per cent is considerably greater than that newly formed by the disintegration of thorium 2. Thus a contamination with traces of the decay products of mesothorium which is unimportant for the beta activity can easily cover the alpha activity which appears with the decay of thorium 2, especially since with the thorium 2, radiothorium is precipitated which for its own part undergoes an increase in activity.

Thus I have not been able to demonstrate with complete certainty the formation of radiothorium or thorium 3 from thorium 2, although I have been able to push down to an extremely low level the alpha activity of the preparations which I have used. Since I have however been able to detect within a few days the formation of radiothorium from mesothorium, that is from Th 1 + Th 2, and since in the present paper I have shown that the thorium 2 which disintegrates in 6.2 hours is a direct disintegration product of thorium 1, the obvious conclusion seems justified that Th 1, Th 2, and Th 3 stand in a direct genetic relationship to one another.

In the following table I give a survey of the first disintegration products of thorium as they are shown by these and earlier findings:

Product	Disintegration period	Nature of radiation
1. Thorium		alpha
2. Thorium 1 (Mesothorium 1)	about 5.5 years	rayless
3. Thorium 2 (Mesothorium 2)	6.20 hours	beta
4. Thorium 3 (Radiothorium)	2 years	alpha
etc.		

Berlin, Chemical Institute of the University

(Received 7 March 1908.)

V. Ionium, the Parent of Radium

[Before Rutherford and Soddy parted company in the spring of 1903, they understood clearly that radium could not be a "permanent" element. With a half-life measured only in thousands of years, it must be younger than the mineral in which it lay,[1] and since radium was found only in uranium ores, it could plausibly be considered to be a transformation product of uranium. When they separated, it was agreed that Soddy would demonstrate this relationship by detecting the growth of radium in uranium.[2]

The experiment was easily planned and should have been straightforward. He would purify a specimen of uranium carefully, leave it in solution for convenience, and seal it up in a flask. From time to time he would extract the dissolved gases from the solution and test them for the continuing ionization which would reveal the presence of the radium emanation. He began in Ramsay's laboratory in London and continued under better conditions when he moved to Glasgow, but the emanation he could detect was always devastatingly less than he expected. In Boltwood's hands, a similar procedure gave even less encouraging results.[3]

Neverthleless, McCoy's equilibrium measurements proved that every radioactive substance in the uranium minerals was a transformation product of uranium, and Boltwood made the proof more pointed when he showed that in every mineral there was a constant ratio of abundance between its uranium and its radium.[4] If uranium did not grow radium

---

[1] E. Rutherford and F. Soddy, "Radioactive Change", *Phil. Mag.* [6], 1903, *5*: 576–591 (Paper 13 in Volume I).

[2] E. Rutherford, "Early Days in Radioactivity", *Journal of the Franklin Institute*, 1924: *198*: 281–289.

[3] F. Soddy, "The Life-History of Radium," *Nature, Lond.*, 1904, *70*: 30; "The Origin of Radium," *ibid.*, 1904–05, *71*: 294; "The Production of Radium from Uranium," *Phil. Mag.* [6], 1905, *9*: 768–779.

B. B. Boltwood, "Production of Radium from Uranium," *Amer. J. Sci.* [4], 1905, *20*: 239–244.

[4] H. N. McCoy, "Ueber das Entstehen des Radiums," *Ber. dtsch chem. Ges.*, 1904, *37*: 2641–2656; "Radioactivity as an Atomic Property," *J. Amer. chem. Soc.*, 1905, *27*: 391–403; "The Relation between the Radioactivity and the Composition of Uranium Compounds," *Phil. Mag.* [6], 1906, *11*: 176–186.

B. B. Boltwood, "Uranium and Radium," *Nature, Lond.*, 1904, *70*: 80; "On the Ratio of Radium to Uranium in Some Minerals," *Amer. J. Sci.* [4], 1904, *18*: 97–103; "The Origin of Radium," *Phil. Mag.* (6), 1905, *9*: 599–613.

E. Rutherford and B. B. Boltwood, "The Relative Proportion of Radium and Uranium in Radio-Active Minerals," *Amer. J. Sci.* (4), 1905, *20*: 55–57; 1906, *22*: 1–3.

directly, one might suspect the existence of a long-lived intermediate product which the necessary purification had removed from the experimental solutions although it had lain naturally with uranium in the minerals.

In the next paper, Boltwood describes his search for that possible parent of radium. In modern terms it is Th^{230}, so it is not surprising that he finally located it in the company of thorium. As a matter of fact, he had noticed a thorium-like, radioactive material early in his work, but he had supposed it to be actinium, which Debierne, its discoverer, had characterized as resembling thorium closely. This was the beginning of a confusion which cost Boltwood some effort to clarify, and for this reason he begins the present paper with Debierne's remarks on the nature of actinium.

Actinium could be recognized by the emanation it released, a radioactive gas with a half-life of about 4 seconds. Boltwood proposed to identify the radium also by an emanation, but since the radium emanation had a half-life of about 4 days, the two were easily distinguished from one another and from the emanation of thorium with its half-life of 1 minute. When Boltwood began the investigation to which this paper is devoted, he supposed that the 4-second emanating substance must be present if the 4-day emanating radium were to grow. It was not at all necessary, as Rutherford pointed out, and once this was clear Boltwood moved directly to the isolation of the radium parent.

Since he believed that it resembled thorium, and since there would be extremely little of it, he deliberately added thorium to his solutions as a "carrier" which would give bulk to the precipitates. In the case of his Solution 2, he used this carrier technique repeatedly in the interests of thorough recovery; with Solution 4 he varied it by employing two carriers, the group of cerite earths which he added to bring down the actinium and the natural thorium of the mineral to carry the parent of radium. When the growth of radium appeared in the thorium fraction, the radium parent had been distinguished from actinium.

From this point, the paper is concerned with experiments of verification and with the exploration of the properties of the new substance, which Boltwood chose to call ionium. In his discussion of its chemical properties, he assigns it an atomic weight of 230 (the weight of radium, 226, plus the weight of the alpha particle emitted by the ionium in its transformation.) This differentiated it from thorium with an atomic weight of 232.5 and with this difference established, Boltwood would assume that eventually ionium and thorium could be separated chemically.

It is also worth noticing that Boltwood identified uranium X, in modern terms Th^{234}, as a substance which accompanied thorium and ionium in his extractions.—A. R.]

17

Bertram B. Boltwood

On Ionium, a New Radio-Active Element

[From *The American Journal of Science* [4], 1908, *25*: 365–381]

In a preliminary paper published about five months ago[5] it was stated that a highly radio-active substance, which differed in general properties from any of the known radio-elements, could be separated by certain chemical operations from uranium minerals. In the present paper it will be shown that the earlier conclusions were correct, and that there occurs in uranium minerals a previously unidentified radio-element having distinctive physical and chemical properties.

Introduction

The discovery of a new radio-element, "actinium," was announced by Debierne in two papers which appeared in the *Comptes rendus de l'Académie des sciences* for October, 1899, and April, 1900.[6] In these papers, which were characterized by the lack of precise experimental details and the absence of explicit statements, the chemical properties of actinium were described as very similar to those of thorium, from which it had not been found possible to separate it completely. The following reactions[7] were given as characteristic for the new substance:

(1) Precipitation in hot solutions, slightly acidified with hydrochloric acid, by an excess of sodium thiosulphate. The active matter is contained almost entirely in the precipitate.

(2) Action of hydrofluoric acid on the freshly precipitated hydrates held in suspension in water. The portion dissolved is only slightly active. By this method titanium may be separated.

(3) Precipitation of neutral nitrate solutions by hydrogen peroxide. The precipitate carries down the active body.

(4) Precipitation of insoluble sulphates. If barium sulphate, for example, is precipitated in the solution containing the active body, the active material is carried down by the barium. The thorium and

---

[5] Boltwood, this Journal, xxiv, 370, 1907; Nature, lxxvi, 544, 589, 1907.

[6] Vol. cxxix, p. 593; vol. cxxx, p. 906.

[7] It will be noted that the first three reactions given are characteristic for thorium.

actinium are freed from barium by converting the sulphate into chloride and precipitating with ammonia.

About two years later, in 1902, Giesel published an account[8] of a highly radio-active substance which he had obtained from pitchblende. He stated that this body was separated with the cerium earths, and could be ultimately removed with the lanthanum. A very characteristic emanation was evolved by Giesel's more concentrated preparations and he at first used the term "emanating substance" to distinguish the active material. The emanating substance was not separated with the thorium, and its chemical deportment was so different from that of the latter element that in Giesel's opinion the "emanating substance" and Debierne's actinium could not be identical. In a later paper[9] the name "emanium" was applied to the "emanating substance."

Giesel's claims for emanium were answered by Debierne in a paper published in 1904,[10] stating that his "actinium" preparations emitted an emanation similar to that noted by Giesel from "emanium" and asserting that the active material contained in them could not therefore be different. To this Giesel replied by calling attention to differences in the observed rates of decay of the active deposits from "emanium" and "actinium," and stating,[11] moreover, that "Debierne no longer uses the thorium of the pitchblende for obtaining the new preparations, but uses instead, as I do, the cerium earths in which the activity is concentrated, as Debierne acknowledges." Further data on the chemical behavior of "emanium" were given in a later paper by Giesel.[12]

Certain experiments described by Marckwald[13] were believed by him to indicate that "actinium" and "emanium" were not identical, but stood in a genetic relationship to one another. A different and correct interpretation of Marckwald's results was, however, given later by Hahn,[14] who showed that they were due to the existence of a product which Hahn named "radio-actinium." It is interesting to note, in passing, that Marckwald states that the thorium precipitated from a solution of rare earths containing "emanium" was at first highly active but had lost its activity some months later.

The question of the similarity of actinium and emanium is discussed by Debierne in the Physikalische Zeitschrift of January 1,

[8] Ber. d. chem. Ges., xxxv, 3608, 1902; ibid., xxxvi, 342, 1903.
[9] Ber. d. chem. Ges., xxxvii, 1696, 1904.
[10] Comptes r., cxxxix, 14, 588, 1904.
[11] Ber. d. chem. Ges., xxxvii, 3963, 1904.
[12] Ber. d. chem. Ges., xxxviii, 775, 1905.
[13] Ber. d. chem. Ges., xxxviii, 2264, 1905.
[14] Ber. d. chem. Ges., xxxix, 1605, 1906; Phil. Mag., xiii, 165, 1907.

1906. This paper adds little to the information contained in earlier publications, but the following quotation[15] is of interest:

"Ich muss hinzufügen, dass ich mich bei allen meinen Versuchen darauf beschränkt habe, nur Präparate zu verwenden, die nach einer gewissen Zeit eine konstante Aktivität behielten. Von diesen Präparate enthielten einige erhebliche Beträge an Thorium. . . . Der Name Aktinium muss für die neue Substanz vorbehalten bleiben, die die Anfangsursache der Radioactivitätserscheinungen ist. Was die Gieselsche Substanz anbelangt, so habe ich schon gezeigt, dass sie mit Aktinium identisch ist, und diese Identität wird jetzt allgemein zugegeben.

"Ich will nochmals daran erinnern, das in meiner ersten Mitteilung über Aktinium (C. r., April, 1900) die Fällung mit seldenen Erden als eine der Eigenschaften der neuen Substanz angegeben ist: . . . [doch] habe ich hinzufügt, 'dass man jedoch nicht versichern könne, ob die aktive Substanz das Thorium bei allen seinen Reaktionen begleiten würde'."[16]

The identity of the emanations evolved by preparations of actinium and emanium has been demonstrated by a number of experimenters. Our knowledge of the physical and chemical properties of the actinium products is due chiefly to the researches of Hahn[17] and Levine.[18]

In the papers published by Hofmann and Zerban[19] much stress was laid on the differences in activity shown by thorium preparations which had been obtained from certain minerals containing various proportions of uranium. Not only were marked differences noted in many instances, but the subsequent behavior of certain of these preparations was quite irregular, some of the preparations losing and others gaining in activity. In some cases what was believed to be an entirely inactive thorium salt was obtained,[20] and the authors reached

---

[15] Phys. Zeitschr., vii, p. 16.

[16] ["I must add that I have confined myself in all my investigations to the use of preparations which maintained a constant activity after a certain time. Several of these preparations contained considerable quantities of thorium. . . . The name of actinium must be reserved for the new substance which is the original cause of the radioactive phenomena. As for Giesel's substance, I have already shown that it is identical with actinium, and this identity is now generally conceded.

I shall recall however that in my first communication on actinium (C. r. April 1900) the precipitation with rare earths is specified as one of the properties of the new substance: . . . yet I added, 'that it cannot yet be ascertained whether the active substance accompanies thorium in all its reactions,'"—trans. by A. R.]

[17] Loc. cit.

[18] Phys. Zeitschr., vii, 513, 812, 1906; ibid., viii, 129, 1907.

[19] Ber. d. chem. Ges., xxxv, 531, 1902; ibid., xxxvi, 3092, 1903.

[20] Also Hofmann and Strauss, Ber. d. chem. Ges., xxxiii, 3126, 1900.

the conclusion that the specific activity of thorium was dependent solely on the amount of uranium with which it was associated, a conclusion which was not justified, as was shown later by the work of Dadourian, McCoy, Eve, and Boltwood.

These excerpts from the literature have been given because they have an interesting bearing on the experiments, which will now be described.

Early Experiments

In the spring of 1899, Mr. Clifford Langley, one of my students in the Sheffield Scientific School, with which I was at that time connected, undertook to repeat under my direction the interesting experiments of M. and Mme. Curie on the separation from pitchblende of the active substances polonium and radium. The work was carried out with about thirty grams of pitchblende, and the products separated were tested for radio-activity in a sheet metal box containing an insulated electrode connected with a quadrant electrometer. Precipitates containing polonium and radium were obtained without difficulty, but on applying the radio-active test to the material from which these substances had been separated we found that the activity of this was apparently greater than was to be expected for the uranium only. After the removal of the uranium by treatment with ammonium sulphide and ammonium carbonate, a small residue remained which proved to be highly active. From a hydrochloric acid solution of this residue the active material was not removed by further treatment with hydrogen sulphide and was not precipitated when the solution was treated with sulphuric acid. On adding an excess of ammonia, a precipitate consisting chiefly of ferric hydroxide was formed. The total active substance present was separated with this precipitate.

The main conclusion reached from these experiments was stated as follows in a thesis presented by Mr. Langley as a candidate for honors in chemistry in June, 1899: "This investigation shows the possibility of the presence of another active element not precipitated by hydrogen sulphide or by sulphuric acid, but which is precipitated by ammonium sulphide and ammonium hydroxide, thus differing from polonium and radium."

Owing to various causes the work was not continued further, until after the announcement by Debierne, in the following October, of the discovery of a new radio-active element which was separated with the iron group and which had chemical properties similar to those of thorium. I then found that the active substance which we had obtained was almost completely precipitated from a dilute hydrochloric acid solution on the addition of an excess of oxalic acid. Further experiments were not practicable because of the very limited amount of material available, but over three years later the substance was again tested and found to have apparently lost none of its original activity. The material was again obtained in hydrochloric acid solution and the active substance was found to be precipitated by treatment with an excess of sodium thiosulphate. This close agreement with the properties of "actinium" given in Debierne's papers appeared to remove all doubt as to the identity of the active substance.

In the winter of 1903–1904 I made a rather systematic investigation of the constituents of a number of radio-active minerals, in the course of which I frequently observed that the rare earths freshly separated from certain primary uraninites, and other similar minerals containing uranium and thorium, always had associated with them a highly radio-active substance. It was further noted that when a solution of the chlorides of the earths was treated with an excess of sodium thiosulphate, the active substance was almost entirely precipitated with the thorium. The material remaining in the filtrate was essentially inactive when first prepared, but gained in activity slowly, until at the end of several months it was nearly as active as the thorium precipitate from which it had been separated. The activity of the thorium preparation remained nearly constant during the same period. It was also found that if a little thorium salt were added to the solution of a mineral initially free from thorium (for example, carnotite) and the thorium afterwards separated in the usual manner, an active substance was obtained with the thorium which appeared to be identical with that obtained in the other cases. If, instead of a thorium salt, a solution of rare earths from cerite was added to the solution of the uranium mineral and these earths were again separated, a highly active body was found to have been removed with them.

It was at first supposed that these results could be explained on the assumption that the active material which was found to accompany the thorium was Debierne's "actinium," while that which accompanied the other rare earths was Giesel's "emanium," a supposition which was somewhat supported by the fact that the production of a short-lived emanation could be detected in the earths in all cases and in the thorium only occasionally. But even this hypothesis did not appear to be defensible in the face of Debierne's emphatic assertions as to the identity of "emanium" and "actinium."

A number of unsuccessful attempts were then made to effect a separation of the thorium from the more strongly active substance which was associated with it. The thorium was repeatedly precipitated in dilute hydrochloric acid solution by an excess of sodium thiosulphate, was precipitated in solutions of the neutral nitrate with hydrogen peroxide, and the hydroxide suspended in water was treated with hydrofluoric acid. The material obtained after this series of operations was to all appearances and chemical tests pure thorium hydroxide, but it still possessed a high activity out of all proportion to the actual amount of thorium present. The possibility of the presence of a new and previously unidentified radio-active substance naturally suggested itself, but in the absence of more convincing evidence the agreement with the chemical behavior of "actinium" as given by Debierne appeared to offer an insurmountable objection to such a conclusion. The numerous radio-active changes which had already been shown to exist in the actinium series were suggestive of the possibility of further complicated relations of a similar nature. The problem was therefore temporarily abandoned. In a paper on the radio-activity of thorium minerals and salts,[21] in discussing the results obtained by Hofmann and Zerban, it was stated that:—"This element (actinium) invariably accompanies the thorium, and in the thorium separated from minerals containing much uranium and little thorium the activity due to the actinium may be much greater than the activity due to the thorium present."

Later Experiments

The definite and constant proportion which has been shown[22] to exist between the quantities of uranium and radium in minerals could be satisfactorily explained only by the assumption that radium was a disintegration product of uranium. On the other hand, attempts to observe the growth of radium in solutions of purified uranium salts had given results[23] which demonstrated that radium was not directly formed from uranium at anything like the rate which was to be expected from the value for the life of radium as calculated by Rutherford. The discrepancy could be accounted for if an intermediate, slowly changing product intervened between uranium and radium, but the existence of such a product had not yet been established. In the search for this product my attention was again directed to the active substances separated with the rare earths from uranium minerals, as the anomalies already mentioned had been observed in these preparations, and as, moreover, my earlier experiments had shown that practically all of the permanent activity, other than that due to uranium, radium and polonium, was associated with this material.

Sometime before this Rutherford had published[24] a brief account

[21] This Journal, xxi, 424, 1906.
[22] Boltwood, Phil. Mag., ix, 603, 1905.
[23] This Journal, xx, 239, 1905.
[24] Phil. Trans. Roy. Soc. Lond., cciv, 218, 1904.

of an experiment in which he had attempted, without success, to detect the growth of radium in a solution of Giesel's "emanating substance," and had stated that measurements extending over a period of three months indicated that if radium was produced at all, it was produced at a very small fraction of the theoretical rate.

In spite of the unpromising outlook, the following experiment was undertaken with the object of further investigating the existing relations. A kilogram of rather low grade carnotite ore was treated for some time with hot, dilute hydrochloric acid and the considerable residue of insoluble matter remaining was removed by filtration. The solution was then treated with hydrogen sulphide and the precipitated sulphides filtered off. The filtrate was heated to boiling, and about two grams of thorium nitrate (from monazite), followed by a large excess of oxalic acid, were added. The mixture was allowed to stand for several days, when the precipitated oxalates were removed, converted into nitrates, and the precipitation with oxalic acid was repeated. The second precipitate, consisting almost wholly of thorium oxalate, was ignited to form the oxide, the oxide was treated with sulphuric acid to obtain the sulphate, and the sulphate in dilute solution was treated with an excess of ammonium hydroxide. The hydroxide was filtered off, washed thoroughly with hot water, dissolved in dilute hydrochloric acid and again precipitated with ammonia. After washing with water, the second precipitate of hydroxides was dissolved in a small volume of hydrochloric acid and the solution[25] was sealed up in a glass bulb. About two months later the gases in the bulb were removed by boiling the solution and the activity of the radium emanation present was determined in an air-tight electroscope. The bulb was then sealed up for a further period of 193 days, when the gases were again collected and tested. The amount of radium emanation found in the second case was nearly twice as great as that found in the first. It was evident that the amount of radium contained in the solution had increased during the interval between the tests and that the particular substance from which radium was derived had been separated with the thorium. In view of the results of my earlier experiments on the activity of the thorium treated in this manner, I was led to assume that the active substance associated with the thorium was actinium, or at all events the "actinium" described by Debierne in his early papers. The results obtained on the growth of radium were published in the form of a preliminary paper[26] and other experiments were immediately begun with the object of more carefully investigating the entire matter.

---

[25] This solution is referred to in the following pages as "solution 1."
[26] Nature, Nov. 15, 1906: this Journal, xxii, 537, 1906.

On learning of the outcome of my experiments, Professor Rutherford then made some further tests of the preparation of Giesel's "emanating substance" which he had earlier examined, and obtained positive evidence of the growth of radium in this material.[27] This tended to strengthen rather than otherwise the presumption that actinium was the intermediate product between uranium and radium, although there appeared to be certain theoretical objections to such a conclusion.[28]

On continuing his experiments still further, Rutherford was able to effect a partial separation of the emanating substance proper from the radium-forming material by treatment of the solution with ammonia and hydrogen sulphide. He therefore concluded that ordinary commercial preparations of actinium contained a new substance which was slowly transformed into radium, was chemically quite distinct from actinium and radium, and was capable of complete separation from them.[29] He was unable, however, to decide whether the new substance emitted ionizing radiations or was rayless. He also obtained results which indicated that the production of radium in solutions of the parent was approximately constant for a period of over 240 days from the time of its preparation.[30]

In the continuation of my own experiments I prepared the following materials:

Solution 2. Another kilogram of the carnotite ore, similar to that used in preparing the first solution, was similarly treated with dilute hydrochloric acid and the insoluble material was separated. This insoluble matter after drying in the air had a weight of about 800 grams and consisted largely of silica. The silica was completely removed by treatment with hydrofluoric and sulphuric acids, and the residue of sulphates remaining was almost entirely soluble in water. The solution of the sulphates was treated with an excess of ammonia and the precipitate of hydroxides was removed and carefully washed with water. The hydroxides were dissolved in dilute hydrochloric acid, the solution was treated with hydrogen sulphide and the precipitated sulphides were filtered off. To the filtrate about one gram of thorium nitrate was first added and then an excess of oxalic acid. The precipitated oxalates were filtered off about two days later. A further quantity of about one gram of thorium nitrate was added to the filtrate and a second precipitation of thorium oxalate was made in the solution. The solution obtained by heating the mineral with hydrochloric acid was worked up by the same series of

---

[27] Nature, lxxv, 270, 1907.
[28] Rutherford, Radioactivity, p. 177.
[29] Nature, lxxvi, 126, 1907.
[30] Phil. Mag., xiv, 733, 1907.

chemical operations as that used in the preparation of the first solution (p. 153). After the precipitate of thorium oxalate had been removed more thorium nitrate was added and a second precipitation of the oxalate was made in the same solution. The object of the second addition of the thorium salt was to insure the removal of any of the new substance which had escaped the first treatment. The four oxalate precipitates were combined, were converted into chlorides and were again treated with oxalic acid. The final oxalate was converted into the chloride, was twice precipitated as the hydroxide to remove the radium as completely as possible, and was finally obtained as the chloride, in which form it was sealed up in a glass bulb and the growth of radium observed. About thirty per cent more radium was produced in this solution in a given time than was produced in the first solution in the same period.

Solution 3. This solution was prepared from a quantity of Joachimsthal pitchblende weighing 200 grams. The chemical operations were essentially the same as those carried out in the preparation of solution 2, except that the mineral was first decomposed with dilute nitric acid. A considerable residue remained after treating the insoluble material with hydrofluoric and sulphuric acids. This residue was obtained in solution by a series of operations, including fusion with sodium carbonate, and this solution was separately treated with thorium. The entire thorium material ultimately recovered was freed from radium as completely as possible by repeated precipitation with ammonia, and the growth of radium was measured in this solution in the usual manner.

Solution 4. This was prepared from about 100 grams of secondary uranium minerals, consisting chiefly of gummite and uranophane. No thorium was introduced into the solution of this material, which contained naturally about one per cent of thorium oxide and about two per cent of other rare earths, but instead there was added a solution of about two grams of oxides of rare earths obtained from a specimen of cerite. This cerite had been found to be wholly free from thorium, uranium or other active substances. The rare earths were then separated in the usual manner, by precipitation as oxalates. The oxalates were converted into chlorides and the thorium was removed in the customary manner by treatment with sodium thiosulphate. The precipitate obtained was digested with hot, dilute hydrochloric acid, the sulphur was filtered off and the treatment with sodium thiosulphate was repeated. The final precipitate was dissolved in hydrochloric acid and the solution was examined for the growth of radium. In 140 days the amount of radium was seven times as great as the amount present at the start.

Solution 4*b*. The rare earths left after the separation of the thorium used for the preparation of solution 4 were recovered by treating the filtrates with an excess of ammonia. They were reprecipitated four times as hydroxides in order to remove radium and, as a solution of the chlorides, were sealed up in a glass bulb. The amount of radium formed in this solution in a period of over 140 days was too small to be detected and could not have been as much as one one-hundredth of the amount formed in solution 4 in the same interval.

Solution 4*c*. The solution from which the thorium and other rare earths used in 4 and 4*b* had been separated was evaporated to dryness and the residue was heated to destroy the oxalic acid present. The residue was dissolved in dilute hydrochloric acid and a solution of about two grams of rare earths from cerite was added. The earths were then separated as before, and, after removal of the radium, were obtained in hydrochloric acid solution. No evidence of the growth of radium in this solution could be obtained. It was apparent that the substance from which the radium was formed had been completely separated with the thorium.

Activity of the Substances in Solutions. After the solutions had been preserved for some time, during which observations of the growth of radium were carried out, the solutions were removed from the bulbs, diluted to a known volume and a definite fraction removed. In this fraction the material present was precipitated as hydroxide with ammonia, and the hydroxide was strongly ignited to form the oxide. From the oxides thus obtained thin films weighing a few milligrams were prepared[31] and the activities of the films were determined in an electroscope. The activities of the films were measured at frequent intervals over a period of about 130 days. The activity of the material from solutions 1, 2, 3 and 4 showed a slight initial rise corresponding to the formation of thorium X in the thorium present, but the final maximum attained was only about two per cent greater than the activity at the start. From the known weights of the films, which had been accurately determined, the total activity of the substances present in each of the solutions could be calculated. On comparing these calculated activities with the amounts of radium produced in the different solutions in equal periods, it was found that these two quantities were closely related to one another. After correcting the activities for the amounts of thorium and thorium products present, it was found that the amount of radium produced in each solution was quite closely proportional to the activity of the other material present. From these results

[31] The films were prepared by the method already described; this Journal xxv, 274, 1908.

it appeared highly probable that the substance from which radium was produced was an element emitting an α radiation. To further test the properties of the active substance the following experiments were performed. Some thorium had been separated from a quantity of uraninite about six months before and had been carefully purified by the well known method of dissolving the oxalate in an excess of ammonium oxalate. The oxalate had been later ignited at a low heat to form the oxide. The activity of this oxide was high and it contained an amount of the new body having an activity about equal to that of five grams of pure uranium. The oxide, weighing 0·4 gram, was placed in a glass tube between plugs of cotton wool, and a strong current of air was drawn though the tube and into a sensitive electroscope. The only emanation which could be detected was that due to thorium.

The greater part of solution 3, described above, was then taken, diluted to a volume of about 800^{cc} and, after heating to boiling, an excess of ammonia was added. The hydroxides were filtered off and the residue obtained on evaporating the filtrate to dryness was examined. The activity of this residue was slight and appeared to be due entirely to thorium X. The hydroxides were dissolved in an excess of hydrochloric acid and a small quantity of sodium thiosulphate was added. The solution was boiled and the separated sulphur was filtered off. It was ignited and the activity of the very slight residue was tested and found to be one-thousandth of that of the total active material in the solution. Some sulphuric acid followed by barium chloride was added to the solution and the precipitate of barium sulphate which formed was removed. The activity of this was tested and was found to be less than the activity of the residue left from the sulphur. It was evident that the active substance in the solution was not actinium, for if it had been the treatment with ammonia would have separated actinium X[32], and the treatment with sodium thiosulphate and barium sulphate would have separated radio-actinium.[33]

A more active preparation of the new substance was then obtained in the following manner: To the solution procured by treating about one kilogram of carnotite ore with dilute hydrochloric acid there was added a few milligrams of thorium nitrate and about five grams of the chlorides of inactive rare earths from cerite. The rare earths were precipitated as oxalates, after the removal of the hydrogen sulphide group, and the thorium was separated from the other rare earths

[32] Giesel, Ber. d. chem. Ges., xxxviii, 775, 1905; Godlewski, Phil. Mag., x, 35, 1905.

[33] Hahn, Phil. Mag., xiii, 165, 1907.

(and the actinium) by repeated precipitation with sodium thiosulphate. In this way a few milligrams of a substance, consisting chiefly of thorium oxide and having an activity at least 1000 times that of an equal weight of uranium oxide, were obtained. The radiations emitted by this material in the form of a thin film were examined in the electroscope and were compared with that emitted by a preparation of polonium on bismuth having an approximately equal activity. It was found that the new body gave out an α radiation which was much more readily absorbed by air and aluminium than the α radiation from polonium. The presence of a β radiation, producing over ten per cent of the total ionization, was also noted. The first measurements were thought to indicate a high absorption, but on comparing the results with those obtained by Levin[34] for uranium X, it was found that the coefficient of absorption for the β rays was the same in both cases.[35] That this β radiation was due to uranium X was also evident from the fact that it fell off exponentially with the time at a rate corresponding to half-value in about 22 days.

An attempt was then made to determine the range of the α particles in air by the scintillation method, and although the difficulties were increased by the presence of a small proportion of α particles of longer range emitted by the thorium products in the material, it was easily ascertained that the range in air of the α particles from the new substance was certainly less than three centimeters. No other radio-active element emitting rays of so short a range had been previously identified. From this and from the data obtained in the other experiments it was evident that the material in hand contained a new radio-active element and the name "ionium" was suggested for this substance.

Separation of Ionium.

As a result of further experiments it was found that highly active preparations of ionium free from appreciable amounts of other radioactive substances could be obtained in the following manner: A quantity of carnotite[36] was treated with hydrochloric acid and several grams of rare earth from cerite in the form of chlorides were added to

---

[34] Phys. Zeit., viii, 585, 1907.

[35] Levin's measurements give the value $17{\cdot}0(\mathrm{cm})^{-1}$ for the coefficient of absorption for thicknesses of aluminium between $0{\cdot}0136$ and $0{\cdot}180^{\mathrm{cm}}$. For the same thicknesses I obtained the same value for the coefficient of absorption of the β rays from my preparation.

[36] The carnotite used in these and the earlier experiments was kindly sent to me by Mr. William Zowe of Uranium, Colo. I have recently obtained some very excellent material of the same sort from Mr. B. J. Manning of Paradox, Colo.

the solution. The earths were then separated by the method already described. The material precipitated from a solution of the earths by treatment with sodium thiosulphate was further purified by repeated precipitation with the same reagent, and the material finally obtained weighed only a little over one milligram. This material had an activity several thousand times greater than that of an equal weight of pure uranium. It contained when first prepared very appreciable amounts of uranium X, but this substance disintegrated with the time until now at the end of about five months its presence can scarcely be detected. Only a relatively small proportion of the total ionium present is obtained by this process.

Range of the α Particles from Ionium.

Dr. L. P. Wheeler of the Sheffield Scientific School and Mr. T. S. Taylor of this Laboratory have both been so kind as to measure for me the range in air of the α particles emitted by ionium. They each used a somewhat modified form of the Bragg apparatus and the measurements were conducted on a small quantity of a very active preparation. Their results, obtained independently, were in excellent agreement and gave a range for the α particles of 2·8 centimeters in air (at 760^{mm} pressure).

Other Radiations from Ionium.

The question as to whether ionium emits also a β and a γ radiation is still an open one, as the preparations which I have are not strong enough or old enough to make it possible as yet to reach a definite conclusion. The indications, however, are in favor of the view that ionium emits β rays which are more readily stopped by matter than the β rays from uranium X.

Relative Activity of Ionium in Minerals.

The results obtained on the determination of the relative activity of the amount of ionium in radio-active equilibrium with uranium and radium in a mineral have already been given in an earlier paper.[37] It was found that the activity of the ionium is about 0·34 and that the activity of the radium is about 0·45; the activity of the uranium with which they are associated being taken as unity. The ratio of the activity of the ionium to the activity of the radium is therefore approximately $\frac{0{\cdot}34}{0{\cdot}45} = 0{\cdot}76$, a value which is in good agreement with

[37] This Journal, xxv, 291, 1908.

the ratio of the ranges of the α particles $\left(\frac{2{\cdot}8}{3{\cdot}5} = 0{\cdot}80\right)$ emitted by the two substances.

Life of Ionium.

Some idea of the probable life of ionium, or, in other words, of the time which would have to elapse before one-half of a given quantity of this element would completely disintegrate into other forms of matter, can be gained from the results of experiments on the production of radium by ionium and on the growths of radium in pure uranium compounds. I have found, for example, that the rate of production of radium in a solution of ionium is constant, within the limits of experimental error, for a period of over 500 days, although a change of the rate by as little as five per cent could probably have been noted with certainty. This would indicate that the half-value period, or time required for half of a given quantity of ionium to be transformed into radium, is at least 25 years. I have also found that the amount of radium formed in a solution containing 48 grams of purified uranium in a period of over two and one-half years is probably less than 10^{-11} gram, and from this result it would appear that the life of ionium is at least as long as that of radium, provided that no other product having a slow rate of change occurs between uranium X and ionium. If the life of ionium is of the same order of magnitude as that of radium, then the amounts of the two substances contained in uranium minerals should be approximately equal. If this is the case, then it ought to be possible ultimately to obtain preparations of pure ionium, just as it has been found possible to obtain pure radium salts. By a comparison of the activity of a known amount of pure ionium with the activity of a definite quantity of pure uranium oxide it would be a simple matter to determine the relative amounts of these elements contained in a mineral.

Ionium in Pitchblende Residues.

Through the kindness of Professor Rutherford and the Royal Society of London it has been possible for the writer to examine a small quantity of material which had been separated from the pitchblende residues presented to the Royal Society by the Austrian Government. The material was in the form of a crude hydroxide containing polonium, actinium, radium, lead, zinc, and a large number of other elements, and weighed, in the form of a wet paste, about 180 grams. It was treated with hot dilute hydrochloric acid, the silica was removed and the substances precipitated by hydrogen sulphide were separated. The filtrate from the sulphides was treated with an excess of oxalic acid, the oxalates were further purified, and treated, as chorides, with an excess of sodium thiosulphate. After repeated precipitation with thiosulphate a few milligrams of material which was free from actinium and actinium products was finally obtained. This material contained ionium having an activity about equal to that of a half gram of pure uranium. The almost complete absence of thorium in this product was quite unexpected, and it is not impossible that the thorium and ionium may have been undesignedly separated in the early treatment of the residue. It is not unlikely, moreover, that the greater proportion of the thorium and ionium present in the original mineral is not retained in the insoluble residue but is removed with the crude uranium salts which are separated in the technical process of manufacture.

Production of Ionium by Uranium.

If ionium is produced directly from uranium X it ought to be possible under suitable conditions to determine this fact by direct experiment. With this object in view I have made some preliminary experiments, using a portion of my carefully recrystallized uranium nitrate[38] prepared nearly four years ago. By certain chemical operations, it was found possible to separate a small quantity of material free from uranium and emitting an α radiation. The final permanent activity of this substance was slight, but was of about the magnitude to be expected for the activity of ionium if this had been formed directly from the uranium. Further experiments on larger quantities of uranium salts are now in progress and it is hoped that the question of the production of ionium by uranium can ultimately be decided with a satisfactory degree of certainty.

Chemical Properties of Ionium.

The chemical properties of pure ionium can not be definitely determined until it has been possible to examine weighable amounts of this element. But from the data which has been already obtained it would appear probable that it belongs to the group of elements commonly known as the rare earths and forms salts which are in general isomorphous with those of thorium. The separation of ionium from thorium presents indeed a difficult problem, and I have been unable to discover any indications that even a partial separation can be effected by the use of such characteristic reactions as the precipitation of the thorium by hydrogen peroxide, sodium thiosulphate, meta-nitro-benzoic acid or fumaric acid. From its position with respect to radium it can be safely assumed that the atomic weight of ionium is probably not far from 230, and the atomic weight of

---

[38] This Journal, xx, 239, 1905.

thorium, 232·5, would bring these two elements into close proximity in the periodic table.

By observations of the growth of radium in purified salts of thorium obtained from monazite by commercial methods, Hahn[39] has demonstrated that ionium is also contained in these compounds.

General Discussion of Results.

It appears highly probable from the data now available that, in the earlier experiments of Debierne, the material which he described under the name of "actinium" consisted in reality of a mixture of ionium with the element first definitely identified and described by Giesel under the name of "emanium." Debierne's statement, quoted in the introduction, that he had used in his experiments only preparations which had retained a constant activity for a certain time, while not very explicit, would at least appear to exclude the possibility that the incorrect chemical properties which he attributed to what is now known as actinium can be explained by any other assumption. That Giesel's preparations of the "emanating substance" have in one case, at least, contained appreciable amounts of ionium, is evident from the results obtained by Rutherford on the growth of radium in a specimen of this material.

The anomalous behavior of the thorium separated from uranium minerals by Hofmann and Zerban can also be explained without difficulty. In those cases where a permanent high activity was shown by the thorium it was probably due to a relatively high proportion of ionium in the preparation. When the freshly separated thorium contained radio-actinium, the activity at first rose and later fell with the time; when actinium itself was present the activity rose only. It is probable that the radio-active measurements were not sufficiently sensitive to detect the activity of thorium and its products when not accompanied by other active substances, so that when thorium was separated from minerals containing but a small proportion of uranium it was erroneously assumed to be inactive.

The existence of ionium also has a bearing on the conclusion reached by Hahn[40] that thorium itself on disintegrating emits an α radiation. Hahn has published as yet no quantitative data in regard to this matter, but measurements which I have carried out by a method similar to that which he has described would appear to indicate that the activity which can be attributed to thorium, from monazite, when free from all thorium disintegration products can not be greater than about seven per cent of the activity of the same

---

[39] Ber. d. chem. Ges., xl, 4415, 1907.
[40] Ibid., xl, 3304, 1907.

thorium when equilibrium amounts of its disintegration products are present. This is so nearly of the same order of magnitude as the relative activity of the ionium undoubtedly associated with the thorium, that there would appear to be some question as to the correctness of Hahn's conclusions.

The results which I have obtained on the rate of growth of radium solutions of ionium, indicating that the half-value period of radium is about 2000 years, will be given in a later paper.

Summary.

It has been found that uranium minerals contain a new radioactive element, to which the name "ionium" has been given. The chemical behavior of ionium is similar to that of thorium, from which it can not be separated by the usual reactions characteristic for thorium.

Ionium emits an α radiation having a range of about $2{\cdot}8^{cm}$ in air, and probably also a β radiation. Results obtained on the growth of radium in solutions of ionium indicate that it is the immediate substance from which radium is formed. It is therefore undoubtedly a disintegration product of uranium intermediate between uranium X and radium. The relative activity of radium and ionium in minerals is in agreement with this assumption.

New Haven, Conn.,
February 28, 1908.

[Marckwald has already appeared in this book as a shrewd and ingenious investigator of polonium. His name was cited by Boltwood in the last paper for his work with actinium. It is not surprising then to find him involved with ionium in the person of his graduate student, Bruno Keetman (1884–1918). The next paper represents Keetman's doctoral dissertation, and indeed owes its comprehensiveness to this academic origin.

Keetman did not find ionium (as his introduction suggests) by following to their logical conclusion the hints laid down by Boltwood, Rutherford, and Hahn. He began, as he himself tells it, with the hope of connecting actinium with radium, and specifically in an attempt to extract substances with a rare-earth chemistry from uranium minerals. What his process yielded was a substance with an unexpectedly intense radioactivity which he recognized as the ionium Boltwood was just then announcing. To be sure that none escaped, he tried to improve his chemistry, and in this way was led to the interesting discovery that no kind of precipitation, crystallization, or sublimation could alter the proportion of ionium to thorium, so alike were they in chemical properties. He also found, as did Boltwood, that the procedures which extracted ionium were equally effective for uranium X.

The competence with which Keetman worked (quite probably with some guidance from Marckwald) appears again in the subsidiary investigations with which the paper closes. There is a determination of the half-life of radium made (with Hahn's help) by ionization measurements, and Keetman's value matches within two per cent the value derived by Rutherford and Geiger from the direct counting of alpha particles. There is an estimate of the half-life of ionium. There is a clear demonstration that ionium does not transform to actinium.

There is also an estimate of the atomic weight of ionium which leads Keetman to the same conclusion as Boltwood, that ionium and thorium owe their close chemical similarity to the small difference in their atomic weights, just as do the rare-earth elements.

The mention of "Bronson resistances" in the last paragraph of the paper may call for an additional comment. H. L. Bronson in Rutherford's laboratory at McGill had devised a substitute for a resistor by mounting two parallel, insulated, metal plates with an air gap between them. The lower plate was coated with polonium and under the proper conditions the ionization current in the gap could be made proportional to the potential difference across it, so that the device operated as an ohmic resistor with a resistance of the order of 10^{11} ohms.[41]—A. R.]

18

Bruno Keetman

On Ionium[42]

[Translation of "Über Ionium," *Jahrbuch der Radioaktivität und Elektronik*, 1909, *6*: 265–274.]

Rutherford's hypothesis of the disintegration of radioactive atoms leads to the conjecture that radium must be formed from uranium. Some years ago experiments were undertaken by F. Soddy[43] as well as by Boltwood[44] to detect the growth of radium from uranium. Although the results were negative at first, Soddy[45] has recently succeeded in establishing the growth of very small quantities of radium in

[41] H. L. Bronson, "Radio-active Measurements by a Constant Deflection Method," *Amer. J. Sci.* [4], 1905, *19*: 185–187.

[42] Extract from the Doctoral Dissertation of the author, Berlin 1909.

[43] Phil. Mag. [6] *9*, 768, ibid. [6] No. 94, 632, 1908.

[44] Amer. Journ. Sc. Silliman [4] *20*, 239.

[45] Phys. Zeitschr. *10*, 396.

completely pure salts of uranium. In any case, these investigations have shown that the transformation proceeds astonishingly more slowly than would be expected from Rutherford's hypothesis for a direct disintegration of uranium into radium. It became necessary therefore to assume the existence of a long-lived intermediate product.

As early as the end of 1906, Boltwood[46] believed he had found that actinium gradually produced radium, and assumed thereupon that this substance was the desired intermediate material. However Rutherford[47] was able to demonstrate that it was not the actinium itself which changed into radium, but that a hitherto unknown substance must be present in the preparation used, which could be separated from actinium by chemical means. At that time he was not able to give a more precise characterization of the radiation and the chemical properties of the new material.

Then there appeared a paper by Boltwood[48] in which it was shown that the new substance, which he called "ionium," emits alpha rays with a range of less than 3 cm, is similar to thorium in its chemical behavior, and produces radium at a constant rate. Boltwood obtained his preparation from carnotite; since its content of thorium was very small, he had to add moderately large amounts of thorium salts in order to separate the ionium-bearing thorium from the mineral by what were hardly characterisitic reactions. For this reason he obtained only weak preparations, mixed with a good deal of thorium which could not be removed.

O. Hahn[49] found that commercial thorium nitrate contained radium, indeed in larger quantities the older it was. This was explained on the grounds that the monazite which served as the starting material contained uranium and therefore ionium as well. Since this element is closely related chemically to thorium, it was present in the thorium salts obtained from the monazite, and was detected there by its transformation into radium.

Before the appearance of this paper, I began at the suggestion of Professor Marckwald to determine the ratio of radium to actinium in various minerals of uranium. The great difficulties which lay in the separation of small quantities of rare earths from the uranium- and iron-bearing solutions induced me to seek new methods of separations, and so I found in September 1907 that all the rare earths are quantitatively precipitated by hydrofluoric acid from strongly acid solutions, even when they are present in extremely small quantity along with

---

[46] Amer. Journ. Sc. *22*, 537.
[47] Phil. Mag. [6] *14*, 733.
[48] Amer. Journ. Sc. *24*, 370, October 1907.
[49] Berichte d. D. chem. Ges. *40*, 4415, 1908.

uranium and iron. To separate the earths from pitchblende by the help of this simple reagent, I dissolved the mineral in nitric acid, fumed it with sulfuric acid, took it up in water, filtered off the radium-barium-lead sulfate, and heated the solution with an excess of hydrofluoric acid. In this way I obtained in place of the expected weakly radiating actinium, a hitherto unknown substance with an astonishingly strong activity; it could be separated from the elements of the cerium and yttrium groups but not from thorium, and released no emanation, thus distinguishing itself from actinium. Further investigation of this substance was anticipated by the paper of Boltwood, referred to above, which appeared in October of that year. It was easy to establish that my substance was identical with ionium.[50]

Ionium is found in all uranium minerals in a constant amount, from which it follows that it is a disintegration product of uranium. In a paper, "On the Radioactivity of Uranium Minerals," Boltwood[51] has given the ratio of the activities of uranium to ionium as 1:0.34.

Since by the employment of hydrofluoric acid as the precipitating agent it was not necessary to add thorium deliberately to the solution, the possibility developed of obtaining strongly active ionium preparations from minerals poor in thorium. To this end, Professor Marckwald and I induced the firm of E. de Haën to extract this material from the residues which had been obtained in working up an American ore of uranium. Five hundred kilograms of the raw material provided about thirty kilograms of fluorides of the rare earths with calcium and uranium which contained some 1/4 per cent of thorium. The experiments to extract this thorium showed the great incompleteness of the known reactions. The customary precipitating agents, oxalic acid in strongly acid solution or sodium thiosulfate, failed, since they were not even approximately quantitative. On the other hand, when an aluminum salt was added and then precipitated with sodium thiosulfate, it was possible to obtain at least the greater part of the ionium-bearing thorium, since the precipitating aluminum hydroxide carried down the thorium. Yet even after repeated applications of this process, 5 to 10 per cent remained in the solution. Finally even this difficulty was overcome by the use of zinc hydroxide suspended in water; this permitted a complete, quantitative precipitation of the ionium and thorium and carried down only a small part of the cerium earths, which can be removed by repeated applications of the reagent. (The fluorides were decomposed by fuming with sulfuric acid.) There was also the further advantage that the zinc hydroxide could be used even in sulfuric acid solution,

---

[50] W. Marckwald and B. Keetman, Ber. d. D. chem. Ges. *41*, 49.
[51] Amer. Journ. Sc. *25*, 269.

while for the thiosulfate precipitation an excess of chlorides was necessary.

In this way 140 grams of ionium-thorium oxalate were obtained, which were some 200 times as active as uranium metal, and whose ionium content represented the quantity in equilibrium with 82 kilograms of uranium or 27 milligrams of radium. By ignition the oxalate could be transformed into the oxide, which was nearly 400 times more active than metallic uranium. This material was used in experiments on the separation of ionium from thorium, and also for the determination of the life of radium.

The quantity of actinium which was contained in the fluorides, and which went toward lanthanum in the most soluble part during the fractional crystallization of the cerite oxalates from hydrochloric acid, was very small. Perhaps it was not present in the raw material, but had already been removed by a chemical reaction (barium sulfate precipitation?).

Even while the procedure for the extraction of the ionium was being worked out, it was observed that in incomplete precipitations it behaved so exactly like thorium that the ratio of the two remained unaltered. Later, systematic investigations confirmed these results, so that any enrichment of the new element appeared to be very difficult. The following attempts at separation were made: fractional precipitation with oxalic acid from acid solution, with hydrochloric acid from ammonium oxalate solution, with ammonia from a phosphate solution, with potassium dichromate, with sodium sulfide, with aniline; crystallization of the double sodium carbonate, of the oxalate from concentrated hydrochloric acid, of the sodium and potassium double sulfates, of the compound formed with acetonyl acetone from benzene solution, and fractional sublimation of this compound in a vacuum. In no case did it happen that any enrichment could be detected.

The alpha rays of ionium have a range of 2.8 cm in air, as Boltwood[52] has found. They are thus less penetrating than those from all the other radioactive elements. Ionium itself does not emit beta rays, but as it is very similar to uranium X in its chemical aspect, and (as far as I have been able to establish up to now) not separable from it, ionium preparations freshly obtained from uranium minerals give a strong beta radiation which decreases by half every 22 days. Thus one can also use the reactions given above for ionium to obtain uranium X from uranium salts.

In the production of strong uranium X preparations from great quantities of uranium salts, the use of hydrofluoric acid is

---

[52] Amer. Journ. Sc. *25,* 365, May 1908.

burdensome. The method mentioned above for obtaining ionium with zinc hydroxide may not be used since the uranium itself will be brought down. On the other hand a separation by the following scheme will succeed: The solution of the uranium salt is strongly acidified, treated with a little sodium phosphate, and cautiously neutralized with dilute caustic soda. Then, at ordinary temperatures, a part of the uranium very gradually separates as the phosphate, and this contains all the uranium X. This small quantity of uranium can then be dissolved in sulfuric acid, and on the addition of a little cerium salt, precipitated with hydrofluoric acid. This method has proved to be very useful for the separation of concentrated preparations of uranium X from great quantities of uranium.

As initially mentioned, Boltwood discovered that ionium transforms into radium. In a later paper[53] this author has published a great number of measurements on different preparations from which he determines the life of radium as some 2000 years. O. Hahn has succeeded in detecting the presence of the radium-producing substance in ordinary thorium, and from this he has arrived at a rough estimate of the half-value constant of radium.[54] It would be of interest to repeat this determination. This can be done with a solution which contains ionium-thorium chloride and whose radium content is determined from time to time by boiling out the emanation present and measuring it in a suitable electroscope. Dr. O. Hahn has had the kindness to carry out these measurements, for which I owe him great thanks. To determine the ionium content of the solution, 1/2500 of it was evaporated on a sheet of aluminum and the activity was measured. It gave a saturation current of 3.42×10^{-12} amp, and thus the whole amount would give 8.55×10^{-9} amp. For 1 gm of uranium, H. N. McCoy and G. C. Ashman[55] found 2.31×10^{-10} amp, which I have confirmed by my own experiments.

According to the work of Boltwood on the relative activity of the separate products in uranium minerals, the activity of the equilibrium quantities comes to uranium : ionium = 1:0.34. Therefore the uranium standing in equilibrium with the ionium present would give a saturation current of $(1 \times 8.55 \times 10^{-9}/0.34) = 2.55 \times 10^{-8}$ amp. Since 1 gm of uranium gives 2.31×10^{-10} amp, the equilibrium quantity of uranium was $(2.52 \times 10^{-8}/2.31 \times 10^{-10}) = 109$ gm. Since 2.94 kg of uranium stand in equilibrium with 1 mgm of radium,[56] the ionium used corresponded to $(0.109/2.94) = 3.71 \times 10^{-2}$ mgm of radium.

---

[53] Amer. Journ. Sc. *25,* 493.
[54] Ber. d. D. Chem. Ges. *40,* 4415.
[55] Amer. Journ. Sc. *26,* 521.
[56] B. B. Boltwood, Amer. Journ. Sc. *25,* 365.

The amounts of radium measured by Dr. Hahn are brought together in the following table:

Date	Days	Quantity of Radium
5/22/08	0	0.48×10^{-6} mgm
6/13/08	22	1.42×10^{-6} ”
7/14/08	53	2.72×10^{-6} ”
8/14/08	84	3.45×10^{-6} ”
11/ 2/08	164	6.75×10^{-6} ”
12/22/08	214	8.60×10^{-6} ”

The following curve also shows the increase.

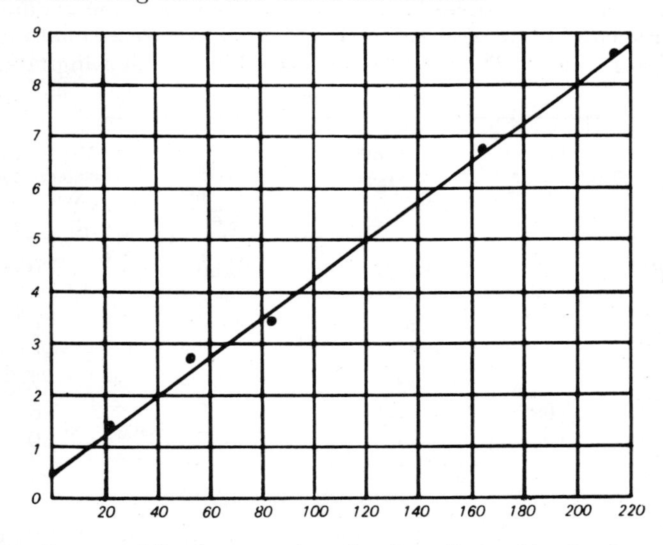

Let Ra_t stand for the quantity of radium formed in the time t, and Ra_∞ for the equilibrium quantity (after an infinite time); then the constant of radioactivity for radium can be calculated from the first and last values in the table as follows:

$$\frac{\text{Ra}_t}{\text{Ra}_\infty} = \frac{(8.60 - 0.48) \times 10^{-6}}{3.71 \times 10^{-2}} = 1 - e^{-\lambda t}.$$

From this the fraction of the radium disintegrating in a year is $\lambda = 3.83 \times 10^{-4}$ and the half-value constant of the radium is

$$T = \frac{\log 2}{\lambda \log e} = 1800 \text{ years}$$

The value recently calculated by Rutherford and Geiger[57] comes to 1760 years.

In the collection of the Institute for Physical Chemistry, where this work was carried out, there are many old uranium preparations, the former work of Rammelsberg. I tried now to detect the presence of ionium in one of them, to arrive at least at an approximate limit for the life of this substance.

The preparation—32 gm of uranyl ammonium carbonate—was dissolved in dilute sulfuric acid, a few milligrams of cerium nitrate were added and then precipitated in a platinum dish with hydrofluoric acid while boiling. The activity of the separated cerium fluoride was measured. Because of its contents of uranium X it possessed a considerable beta radiation, but on long-continued measurements it was noticeable that there was also a small alpha radiation present. The results are collected in the following table:

Date	Time in days	Per cent activity (obs.)	Per cent activity calc. for UX	Difference between calc. and obs. activity
1/30/08	0	100.0	100.0	0.0
2/15	16	65.3	60.4	4.9
2/21	22	56.0	50.0	6.0
3/2	32	41.5	36.5	5.0
3/5	35	39.8	33.1	6.7
4/28	89	13.2	7.6	5.6
8/5	188	5.5	0.55	5.0
11/6	281	4.7	0.04	4.7

From the fifth column, one can see that in addition to the regularly decaying radiation of uranium X there was present a constant activity (within the limits of experimental error). To show that this radiation actually came from the ionium, at the conclusion of the measurements the preparation was covered with aluminum leaf of about 0.004 mm thickness and measured again. The current was reduced to 34 per cent of its original value. The absorptive power of the same aluminum leaf was tested on an ionium preparation and this also showed a reduction to 34 per cent. There is no doubt that ionium was present.

Since the age of the preparation could not be established for

[57] Phys. Zeitschr. *10,* 46, 1909.

certain, it can be concluded from this experiment only that the half-value constant for ionium comes to several thousand years. This result agrees with that from Soddy's experiments on the production of radium from uranium which have already been mentioned.

At present experiments are in progress with three other beautifully crystallized uranium salts of approximately known age. Since the separation of ionium from uranium is achieved so easily, it can be assumed that these salts were originally free from ionium. In this case, in all the preparations equal quantities of ionium must be found with the same quantity of uranium.

The question whether the formation of ionium can be established within a few months in great quantities of pure uranium salts is now under attack.

Concerning the origin of the actinium, great uncertainty rules. Boltwood showed that it is probably present in constant amount in uranium minerals[58] and is therefore a disintegration product of uranium, but where it branches from the family tree of radium is completely unknown.

With the ionium material at my disposal, I was able to decide the question whether actinium is a direct transformation product of ionium. 50 gm of oxalate which corresponded to the equilibrium quantity from approximately 30 kg of uranium were diluted, and filtered after several days. The undissolved material must have contained the yttrium earths and the actinium. After decomposition by concentrated sulfuric acid, it was taken up with water, neutralized, and the ionium and thorium precipitated by zinc hydroxide. The yttrium earths were removed from the filtrate with hydrofluoric acid. They had a moderately strong activity which came chiefly from thorium X. After this had decayed for the most part, there still remained a constant activity. An absorption test showed that ionium was not present in noticeable quantity, since an aluminum leaf decreased the radiation to 50 per cent (ionium goes to 35 per cent). The substance was then ignited for a short time, by which the activity was strongly decreased and sank still farther within a few hours. Thus radium was present, and closer calculation showed that this and the remainder of the thorium X accounted for the whole activity. As already stated, a very great quantity of ionium was used; it can therefore be assumed with considerable certainty that actinium is not a direct disintegration product of ionium. It must then possess a life of at least a few hundred thousand years.

The uranium atom, after the loss of two alpha particles,[59] trans-

---

[58] Amer. Journ. Sc. *25,* 269.
[59] Boltwood, Amer. Journ. Sc. *25,* 269

forms into uranium X. This probably turns directly into ionium with the emission of an electron, and the latter transforms into radium. According to Rutherford's latest determination,[60] the atomic weight of the alpha particle is about 3.9; then the atomic weights of the three elements mentioned are calculated as follows:

$$\begin{array}{ccccc} \mathrm{U} & - & 2\alpha & \rightarrow & \mathrm{UX} \\ 238.5 & - & 2 \times 3.9 & = & 230.7 \end{array}$$

$$\begin{array}{ccccc} \mathrm{UX} & - & \beta & \rightarrow & \mathrm{Io} \\ 230.7 & - & 0.0 & = & 230.7 \end{array}$$

$$\begin{array}{ccccc} \mathrm{Io} & - & \alpha & \rightarrow & \mathrm{Ra} \\ 230.7 & - & 3.9 & = & 226.8 \text{ (experimental 226.4)} \end{array}$$

The value calculated here for the atomic weight of uranium X and ionium (230.7) is close to that of thorium (232.5), and it is probable that they belong in the same place in the periodic system, whence the great similarity of the three elements obviously appears. Analogous behavior occurs with neodymium (143.6) and praseodymium (140.5), which belong of course to the group of the rare earths and are very difficult to separate.

According to my investigations, the uranium residues from Joachimsthal contain only about 20 per cent of the equilibrium quantity of ionium. Boltwood found even less there,[61] but his method of extraction (oxalic acid and thiosulfate) was certainly not quantitative. The chief part of the ionium must go into the uranium liquor, and should perhaps be sought in the residues of the uranium manufacture.

The firm of E. de Haën, at Seelze near Hannover, is currently bringing onto the market an ionium-thorium oxalate prepared according to my directions. Because of its strong and very constant activity it is suitable for comparative measurements, e.g. to test electrometers and their associated apparatus for constancy and to determine capacities. Furthermore, this is an excellent ionizer for Bronson resistances, for which polonium with its relatively rapid decay has been used up to now. The employment of radium has the disadvantage that the apparatus must be entirely enclosed because of the emanation, and the resistance therefore can be altered only with difficulty. With the use of ionium however, the resistance can be raised to any suitable value by simple rarefaction of the air without at the same time changing the capacity. It is best to convert the

---

[60] Phys. Zeitschr. *10*, 46, 1909.
[61] Amer. Journ. Sc. *25*, 365.

ionium into the acetonyl acetone compound, dissolve this in chloroform, and let the solution evaporate on the two condenser plates. One should not use more than 0.1 gm to 100 sq cm of surface, otherwise too much radiation is absorbed.

Institute for Physical Chemistry of the University of Berlin

(Received July 27, 1909)

VI. Mesothorium and Some Valuable Clues

[Through Keetman's work, Marckwald was personally acquainted with the chemical resemblance to thorium which had been observed in ionium, radiothorium, and uranium X. He seems then to have been more pleased than surprised to find a similar resemblance between mesothorium 1 and radium and to notice the emergence of a general pattern of similarity-groups in radiochemistry. Indeed he appears to have been as much concerned over the commercial dangers of the situation as with the novelty of his discovery.

It is interesting that he speaks of radium and mesothorium 1 both as "radioactive elements" and as "metabolons." *Metabolon* was a word which Rutherford and Soddy had coined in 1903 as a name for the "fragments of atoms or new atoms" which must make up the short-lived radioactive substances they were investigating.[1] It implied impermanence at the very least, perhaps indeed a fundamental difference between these radioactive substances and the accepted elements. The new term had not proved useful; even its inventors had instantly abandoned it. Its appearance here, seven years later, may indicate that Marckwald had reservations about the fundamental nature of substances which were so easily recognized by their radioactivity, but so difficult to distinguish by their chemical behavior.—A. R.]

---

[1] E. Rutherford and F. Soddy, "Radioactive Change," *Phil. Mag.* [6], 1903, *5*: 576–591 (Reprinted as Paper 13 in Vol. I.)

19

W. Marckwald

Toward the Knowledge of Mesothorium

[Translation of "Zur Kenntnis des Mesothoriums," *Berichte der deutschen chemischen Gesellschaft*, 1910, *43*: 3420–3422.]

(From the Institute for Physical Chemistry of the University of Berlin.)

(Received 28 November 1910.)

In the year 1907 O. Hahn[2] first demonstrated indirectly that in its atomic disintegration thorium transformed itself into a metabolon whose half-value constant came to some $5\frac{1}{2}$ years and which was separated from the thorium in the process of manufacture during its technical extraction from the ore. He called it mesothorium. Later Hahn found this substance in the manufacturing residues. He showed[3] that mesothorium was not directly transformed into radiothorium, but that a short-lived, intermediate product, mesothorium II, was formed (half-value constant 6.2 hours.) More recently[4] Hahn has succeeded in concentrating the material so much that its efficacy, weight for weight, exceeds many times that of pure radium salts. Concerning the chemical properties of mesothorium I and the manner of its detection and extraction the author has published nothing.

Recently a "radium preparation" was delivered to me for study by a chemical manufacturer so that I might determine its radium content. It consisted chiefly of barium chloride. According to its gamma radiation it must have had a content of more than 1% of radium chloride. When its emanating power was investigated, however, the radium emanation it evolved corresponded to a content of only 0.2% radium. Closer investigations showed that the greatest part of the gamma radiation (some 80%) arose from mesothorium II. That is to say that when an aqueous solution of the salt was treated with a trace of iron chloride and then the solution made ammoniacal, the mesothorium II precipitated with the iron hydroxide. From

---

[2] This journal *40*, 1462, 3304 (1907).

[3] Ztschr. f. physikal. Chem. *9*, 246 (1908).

[4] Lecture at the Session of the German Chemical Society, 11 July of this year.

the precipitate a strong gamma radiation was given out whereas the barium chloride, recovered by evaporating the solution, had lost almost its entire gamma radiation. While the precipitate lost its radiation with a half-value constant of something over 6 hours, the salt recovered the greatest part of its radiation within a few days. The ammoniacal precipitate should also contain radiothorium formed from the mesothorium. In fact after the decay of the mesothorium II there remained an alpha-ray residue which yielded thorium emanation.

On inquiry the manufacturer confirmed that his "radium preparation" had been obtained by processing the residues of uranium- and thorium-bearing ores.

Obviously mesothorium is chemically very like radium. I have not yet succeeded in finding a reaction to separate it from barium and radium. On crystallization of the chloride it is enriched with the radium in the more difficultly soluble crystals.

This complete similarity of the two radioactive elements is very interesting. It has not yet been possible to separate the four elements thorium, radiothorium, ionium and uranium X by chemical reactions. It appears now that a second group of radioactive elements exists whose chemical properties are the same, to which, in addition to radium and mesothorium perhaps other metabolons belong.

From the knowledge of the chemical properties of mesothorium, the process of its extraction from the residues of thorium manufacture is given. It is simply to be copied from that for the extraction of radium from the uranium residues. To be sure, because of its uranium content, mesothorium preparations extracted from monazite would show also some content of radium so long as it remains impossible to separate these two substances. On the other hand, radium derived from uranium-rich thorium ores like thorianite will always contain mesothorium.

Since radium has something like 300 times the life of mesothorium, it is of the greatest importance in the purchase of "radium preparations" to consider the possibility of some mesothorium content. Though in the case I have investigated an inferior preparation was put on the market in good faith, the danger of an intentional falsification is obvious. At present the simplest means of testing a radium preparation for a possible content of mesothorium would be to heat it for a short time in order to drive off the emanation. The preparation must then lose its gamma radiation in the course of a few hours, and this will reach full value again in the course of several weeks. Instead of heating it, one can of course dissolve it in water and evaporate the solution. If any gamma radiation remains after the driving off of the emanation and the subsequent decay in a few hours of the radium C, it comes from mesothorium. The ratio of the gamma radiation before and after the treatment gives a measure of the proportions of radium and mesothorium.

[The next paper presents Frederick Soddy's discovery, independent of Marckwald's and almost exactly simultaneous with it, that mesothorium 1 and radium were chemically inseparable. Simultaneous discoveries are not necessarily of equal value; these two were certainly not. Where Marckwald saw a problem in radioactive assaying, Soddy seized upon a new idea concerning the nature of elements. He saw that "chemically inseparable" might also mean "chemically identical" and thus that substances which were easy to distinguish by their radiations and half-lives, and which quite probably had different atomic weights, might among them all constitute only a single element. In this paper, completed at the close of 1910, he proposed on somewhat less than adequate evidence a fully realized hypothesis of isotopes.

The sources of his insight are obscure. Since leaving Montreal in the spring of 1903, he had struggled fruitlessly with the connection between uranium and radium. He had first sealed away a solution of purified uranium in Ramsay's laboratory in London to wait for the growth of radium in it, and he had continued these experiments when he moved to Glasgow in the fall of 1904.[5] After Boltwood's discovery of ionium,[6] he turned partly to a different line, trying to identify the substance into which uranium X transformed. This substance might or might not have been the parent of radium (it was two years before Soddy could bring himself to use Boltwood's name of *ionium*[7]), but here again he was dis-

---

[5] F. Soddy, "Life History of Radium," *Nature, Lond.*, 1904, *70*: 30; "The Origin of Radium," *Nature, Lond.*, 1904–5, *71*: 294; "The Production of Radium from Uranium," *Phil. Mag.* [6], 1905, *9*: 768–779 [Soddy later counted this paper as No. I in a numbered series on this topic.—A. R.]; "The Origin of Radium," *Nature, Lond.*, 1907, *76*: 150.

F. Soddy and T. D. Mackenzie, "The Relation Between Uranium and Radium," *Phil. Mag.* [6], 1907, *14*: 272–295. [This was later counted as No. II in the series.—A.R.]

F. Soddy, "The Relation Between Uranium and Radium III," *Phil. Mag.* [6], 1909, *80*: 308–309; "Die Bildung von Radium aus Uranium," *Phys. Z.*, 1909, *10*: 396; "The Relation Between Uranium and Radium IV, "*Phil. Mag.* [6], 1909, *18*: 846–858; "The Relation Between Uranium and Radium V," *ibid.*, 1910, *20*: 342–345; "Essais pour évaluer la période de l'ionium," *Radium, Paris*, 1910, *7*: 295–300.

[6] B. B. Boltwood, "On Ionium, a New Radio-Active Element," *Amer. J. Sci.* [4], 1908, *25*: 365–381 (Reprinted as Paper 17 in Vol. 2).

[7] [Soddy's first use of the word *ionium* occurs in the group of three papers printed together in Vol. 20 of the *Phil. Mag.* (August, 1910, listed here under footnotes 5, 8, 10. Previously he had spoken of "the parent of radium."—A. R.]

couraged at finding no well-defined chemical properties for uranium X.[8] As minor digressions, he had attempted to measure the production of helium by thorium and uranium,[9] he had considered the relation between the actinium series and the radium series of radio-elements,[10] and with his students had recently made studies on the ratio of uranium to radium in minerals.[11]

Nevertheless, in spite of his failure to make substantial progress with his chosen problems, Soddy knew the radioactivity of the long-period substances near the beginning of the uranium and thorium chains. The closing paragraphs of this paper, in which he explores the consequences of chemical identity, are remarkable both for the correctness with which he applies his new idea to what was already known and for the breadth of his speculation beyond that. Even here at the beginning, Soddy is already willing to abandon atomic weight as the distinguishing property of an element. He recognizes the existence of three isotopes each of thorium and radium and two of lead. (The chemistry of Pb^{210}, radio-lead, had been established by B. Szilard and H. Herchfinkel, two Eastern European visitors at Marie Curie's laboratory.[12]) He recognizes the existence of a pair of isobars between thorium and radium. He suggests that there may be non-radioactive isotopes. He suggests the existence of U^{234}. After his own difficulties with that substance, he could not quite admit uranium X as a thorium isotope, and yet, with his basic idea triumphing over his prejudices, in his last footnote he explains a curious observation from the literature precisely by the isotopic identity of thorium and uranium X.—A. R.]

---

[8] F. Soddy, "The Product and Rays of Uranium X," *Nature, Lond.*, 1908–09, *79*: 366–367; "The Rays of Uranium X," *ibid.*, 1909, *80*: 37–38.

F. Soddy and A. S. Russell, "γ-Rays of Uranium," *Nature, Lond.* 1909, *80*: 7–8; "The γ-Rays of Uranium and Radium," *Phil. Mag.* [6], 1909, *18*: 620–649; "Decay Constant of Uranium X," *ibid.*, 1910, *19*: 847–851.

F. Soddy, "The Rays and Products of Uranium X," *Phil. Mag.* [6], 1909, *18*: 858–865, also *Proc. phys. Soc. Lond.* 1909–10, *22*: 57–67; "The Rays and Products of Uranium X," *Phil. Mag.* [6], 1910, *20*: 342–345.

[9] F. Soddy, "Attempts to Detect the Production of Helium from the Primary Radio-elements," *Phil. Mag.* [6], 1908, *16*: 513–530; "Production of Helium from Uranium," *Nature, Lond.*, 1908–09, *79*: 129.

[10] F. Soddy, "Multiple Atomic Disintegration. A Suggestion in Radioactive Theory," *Phil. Mag.* [6], 1909, *18*: 739–744.

[11] F. Soddy and R. Pirret, "The Ratio Between Uranium and Radium in Minerals," *Phil. Mag.* [6], 1910, *20*: 345–349.

A. S. Russell, "The Ratio Between Uranium and Radium in Minerals," *Nature, Lond.*, 1910, *84*: 238–239.

F. Soddy, "Ratio Between Uranium and Radium in Minerals," *Nature, Lond.* 1910, *84*: 296–297.

R. Pirret and F. Soddy, "The Ratio Between Uranium and Radium II," *Phil. Mag.* [6], 1911, *21*: 652–658.

[12] B. Szilard, "Étude sur le radioplomb," *C. R. Acad. Sci., Paris*, 1908, *146:* 116–118; "Étude sur le radioplomb," *Radium, Paris*, 1908, *5*: 1–5.

H. Herchfinkel, "Sur le radioplomb," *Radium, Paris*, 1910, *7*: 198–200.

20

Frederick Soddy

The Chemistry of Mesothorium[13]

[From *Journal of the Chemical Society*, 1911, *99*: 72–83]

Mesothorium, the first product of the thorium disintegration series, and the parent of radiothorium, was discovered three years ago by O. Hahn (*Ber.*, 1907, **40**, 1462). Although for the past few months powerfully radioactive preparations of this substance, prepared by Hahn during the course of the manufacture of thorium salts from monazite sand in the works of Dr. Knöfler & Co., Plotzensee, Berlin, have been put on the market, nothing has been allowed to transpire concerning the chemical processes by which the separation is effected. Rutherford, in a Royal Institution Lecture (January 31st, 1908), drew attention to the value of mesothorium as a possible substitute for radium in many experiments, owing to the large scale on which thorium compounds are commercially manufactured. Although mesothorium is not a permanent source of radioactivity, like radium practically is, yet the period of average life, about eight years, is sufficiently long to make it a very useful substance if it could be prepared on a large scale. At the beginning of the present year, powerful sources of thorium radioactivity were needed for certain experiments, and as at that time there seemed no possibility of obtaining it in any other way, experiments were undertaken on the separation of mesothorium from thorium minerals, commencing with thorianite.

The discovery of radiothorium by Sir William Ramsay and O. Hahn, which preceded that of mesothorium, was made in the course of the treatment of a large quantity of thorianite for radium. The material was fused with potassium hydrogen sulphate, the insoluble residue being treated in the same way as the radium-containing residues from pitchblende, to extract the barium as chloride. Radiothorium was discovered in this product (Sir W. Ramsay, *J. Chim. Phys.*, 1905, **3**, 617; O. Hahn, *Jahrb. Radioactiv. Elektronik.*, 1905, **2**, 234). Subsequently a study of the chemical properties of radiothorium by Elster and Geitel (*Physikal. Zeitsch.*, 1906, **7**, 445) and G. A. Blanc (*Phil. Mag.*, 1905, [vi], **9**, 148), and the work of Boltwood

[13] Since this paper was written there has appeared a communication on the same subject by W. Marckwald (*Ber.*, 1910, **43**, 3420) which anticipates the discovery of the chemical identity of mesothorium-1 and radium.

(*Amer. J. Sci.*, 1907, [iv], **25**, 93), resulted in the view at present held, that radiothorium resembles thorium completely and cannot be separated from it by chemical processes, and that the radiothorium prepared by Hahn was due to the separation of mesothorium and the spontaneous production of radiothorium from it with the lapse of time. What little is known of the chemistry of mesothorium we owe to Boltwood, who showed that the original process employed in the separation of thorium-*X* (Rutherford and Soddy, Trans., 1902, **81**, 343, 837) in the filtrate after precipitation of thorium by ammonia, separated also the mesothorium. He also precipitated barium sulphate in a dilute solution of a thorium salt, thinking that the absorption of radiothorium by the barium sulphate might explain Hahn's results with thorianite. He found, however, that mesothorium, but not radiothorium, was separated with the barium sulphate, and concludes: "It appears quite likely therefore that the entraining action of barium sulphate on mesothorium was directly responsible for the presence of radiothorium in the thorianite residues [of Hahn]*." Since Sir W. Ramsay and Hahn were dealing with a mixture of radiothorium and mesothorium in genetic relationship, nothing can be learnt from the original papers on radiothorium (*loc. cit.*) of the chemistry of mesothorium,[14] as it is uncertain throughout to which substance the reactions apply, whilst in Hahn's later papers on mesothorium (*loc. cit.*, and *Physikal. Zeitsch.*, 1907, **8**, 277) no information on this subject is given.

Thorianite is, from the radioactive point of view, the most complex material it is possible to work with, as it contains every one of the thirty or more radioactive elements known, in important quantity, the penetrating radiation contributed by the uranium and thorium series being, for the specimen of thorianite examined (which contained about three times as much thorium as uranium), of very similar intensity. The thorium disintegration series, as at present known, is shown below to facilitate reference, and no phenomena were encountered in the course of the work which could not readily be explained by this succession of changes and by the known changes in the radium series.

Throughout, γ-ray methods of measurement with lead electro-

---

* [The square brackets are Soddy's.—A.R.]

[14] Thus Sir W. Ramsay, for example, in his paper "Radio-thorium" (*J. Chim. Phys.*, 1905, **3**, 623) says, "Le nouveau corps se précipite partiellement avec le radium, en ajoutant aux sels de thorium d'abord un sel de baryum et ensuite de l'acide sulfurique; et l'on peut effectuer sa séparation du radium par un des procédés dont nous avons donné la description." [The new substance is partially precipitated with radium on the addition to the thorium salts first of a salt of barium and then of sulfuric acid; and its separation from the radium can be accomplished by one of the procedures whose description we have given.—trans. by A. R.]

scopes (Soddy and Russell, *Phil. Mag.*, 1910, [vi], **19**, 752) have been employed whenever possible, as by these means the difficulty of comparing together specimens differing greatly in density and weight is avoided. The only objection to it is that considerable amounts of material must be employed to produce effects sufficiently large to measure with accuracy. In dealing with monazite, a special very large lead cylindrical electroscope (22 cm. high and 22 cm. diameter) was constructed so as to give greater sensitiveness. The lead used, which was 3 to 4 mm. thick, was taken from the roof of a very old building, in which any radio-lead initially present would have decayed (*Ann. Reports*, 1906, **3**, 365), so that the natural leak might be as small as possible. It proved very successful. A very complete

Thorium. Over 10^{10} years. → Mesothorium-1. 8 years. → Mesothorium-2. 9 hours. → Radiothorium. 2·9 years. → Thorium-*X*. 5·8 days. → Thorium emanation. 77 seconds. → Thorium-*A*. 88 minutes. → Thorium-*B*. 15·4 hours. → Thorium-*C*. ? (very short). → Thorium-*D*. 4·5 minutes. → ?

study of the γ-radiation of the various radio-elements, in conjunction with Mr. Russell, has recently been concluded, and the section dealing with the γ-rays of thorium has appeared in the *Philosophical Magazine* (1911, [vi], **21**, 130). It suffices to state here that the two thorium types of γ-rays, from mesothorium-2 and thorium-*D*, both in penetrating power and in the ratio of their intensity to that of the accompanying β-rays, are extremely similar to the radium type of γ-rays which are given by radium-*C*. Radium-*C*, mesothorium-2, and thorium-*D*, alone of all the known radioactive constituents, give γ-radiation of sufficient intensity to affect the measurements. When β-ray methods are employed, the possible effects due to uranium-*X*, radium-*E*, and, perhaps, actinium-*C* must also be remembered, but the γ-rays from these substances are negligible. In all measurements of γ-rays, the preparations were contained in sealed test-tubes, so as to retain the radium emanation completely. In spite of the complexity of thorianite, no difficulty was ever experienced in deducing from the variations of the γ-activity with time the radioactive constituents present and their relative amount, although two of the periods (those of the radium emanation and thorium-*X* controlling the γ-ray variations are identical.

For the specimen of thorianite worked with, moreover, the γ-activity contributed by thorium-*D* was very similar in intensity to that contributed by radium-*C*, so that, frequently, preparations appeared to remain of constant activity for long periods, when what was really taking place was a simultaneous complete decay of thorium-*X* (and in consequence thorium-*D*) and a concomitant reproduction of radium emanation (and in consequence radium-*C*) from the radium. The analysis of the effects was no doubt enormously simplified, because, as will appear in the sequel, mesothorium-1, radium, and thorium-*X* appear to form a trio of chemically non-separable elements.

Experiments were started with a solution of thorianite in nitric acid by the two known methods, capable of separating mesothorium, due to Boltwood. The repeated precipitation of the solution with ammonia and evaporation and ignition of the filtrate, as in the preparation of thorium-*X*, effects a complete separation in sufficiently dilute solution, but in an attempt to apply it to considerable quantities with more concentrated solutions, the separation was partial, and the fraction separated varied capriciously. The precipitation in the thorianite solution of barium as sulphate, although at first not very successful, was found ultimately, under proper conditions, to yield good results, even in concentrated solutions the separation being nearly complete. In one experiment 26 grams of barium nitrate was dissolved in a solution of 600 grams of thorianite in nitric acid, from which most of the excess of acid had been evaporated. The solution, of volume about 1·5 litres, was put in a Winchester quart bottle, excess of dilute sulphuric acid added, and the bottle shaken for an hour on a shaking machine. The separation of the mesothorium was practically complete, a second precipitation yielding a precipitate possessing an activity not greater than could be accounted for by the regeneration of thorium-*X* in the time between the precipitations.

About 5 kilograms of thorianite were treated by this method, and the γ-activity of the products, sealed up in test-tubes, was kept under observation for some time. The first precipitate from the solution increased rapidly in γ-activity by about 50 per cent. in the course of two days, and then remained nearly constant. During filtering and drying, mesothorium-2 is usually produced in nearly equilibrium amount before the first measurement. The increase for the first two days is due to the formation of thorium-*D* from thorium-*X*. After that time the decay of thorium-*X* just balanced the production of radium-*C* from the radium as already explained, so that the activity remains sensibly constant. For the subsequent precipitates the γ-activity was due mainly to thorium-*X*, regenerated between the precipitations, and, after reaching the maximum, steadily decayed.

In this way, from the 5 kilograms of thorianite a total weight of about 200 grams of barium sulphates of various activities was separated. Up to this time it was considered, as Boltwood had supposed, that the mesothorium was merely absorbed by the barium sulphate. Uranium-*X* was, in a recent research (F. Soddy and A. S. Russell, *Phil. Mag.*, 1909 [vi], **18**, 620), frequently separated by absorption with barium sulphate, and was found to be readily separable from the barium after solution of the barium sulphate, by precipitating with ammonia in presence of a trace of iron.

In the hope of being able to effect a preliminary concentration of the mesothorium from the barium, to a portion of one of the sulphate precipitates, after conversion into chloride, a small quantity of dilute sulphuric acid was added to precipitate a small fraction of the barium as sulphate. The precipitate so obtained, however, was no more and no less active than the same weight of original material. Other attempts to concentrate the mesothorium from the barium, chemically, failed, and the conclusion was drawn that the separation of mesothorium with barium is due, not to an adsorption by the barium sulphate, as had previously been assumed, but to a chemical resemblance between the two elements.

The whole of the active barium sulphate was converted into chloride by ignition with sugar-carbon in separate small quantities in quartz crucibles, and the barium sulphide dissolved in hydrochloric acid. The crude chlorides in acid solution were freed from lead by hydrogen sulphide, made alkaline with ammonia, and filtered from the small iron and thorium hydroxide precipitate. A fractional crystallisation of the barium chloride was then commenced to concentrate the radium present from the barium. The various precipitates had been sorted into three grades, according to their activity, and worked up into pure chlorides separately. The raw material for the fractionations thus comprised three preparations of weights 52, 69, and 105 grams, and relative activities roughly as 5:2:1. As it was then unknown whether the separation of the barium from the mesothorium would be possible, the fractions were not mixed, but dissolved separately. Fractionation was carried out for separating radium from barium in the usual way, the mother liquor of the richer fraction being used to dissolve the crystals of the next richest. Preliminary tests showed that the process was very effective in concentrating the mesothorium as well as the radium from the barium. In a few days three fractions were again made up, of weights 57, 97·5, and 76·5 grams. Mesothorium-2 was separated from each separately by adding 10 milligrams of thorium nitrate, and precipitating with ammonia. The relative activities of these mesothorium-2 precipitates were as 25·3:4·3:1. The mesothorium in equal weights of barium was therefore as 34:3·4:1. This showed that the mesothorium had been effectively concentrated by the fractional crystallisation, the crystals being enriched, and the mother liquor impoverished, as in the case of radium. The three fractions were dissolved, left for a week, then evaporated to dryness and sealed up in test-tubes, and the γ-activity measured over a period of six weeks, 10 grams only of the first fraction being taken for the test. The ratio between the initial and subsequently generated activity gives the relative proportions of mesothorium and radium. These ratios were in order of richness of the preparation, 0·52, 0·46, and 0·56. Hence mesothorium follows the radium extremely closely in the fractionations. The small differences in the ratio, since the raw material of the fractionations was not homogeneous, but derived from different quantities of thorianite, are not greater that might be accounted for by variations in the composition of the mineral, or by the presence, possibly, of thorium-*X* from regenerated radiothorium. The measurements were only rough, whereas those which follow were done with the greatest possible accuracy.

In order to settle whether any alteration at all in the proportion of the mesothorium and radium was produced by fractional crystallisation, the most active fraction, all but the 10 gram sample sealed up for the previous tests, was refractionated as before. The richest fractions, withdrawn from the process after the fifth and eighth successive fractionations, were combined and labelled *AA*. A part of the 10 gram sample of the original material was taken, and labelled *A*. The two specimens were dissolved in water at the same time, freed from mesothorium-2 by precipitation of thorium hydroxide in the solution as before, evaporated, and sealed up. *A* weighed 2·38 grams, and *AA* 2·09 grams. The γ-rays were compared after two days, when mesothorium-2 is again in equilibrium, and at intervals subsequently. The measurements give the means of telling exactly whether any alteration in the ratio of the radium and mesothorium has been effected by the further fractionation.

Fraction *AA* proved to be 8·75 times as active as fraction *A*, showing that a concentration of the active material in the ratio 10:1 had been effected. Nevertheless, the proportions of radium and mesothorium in the two preparations were identical. The ratio of the activities remained unchanged within the error of measurement, which may be estimated at less than 2 per cent., over the period from the first day after preparation onward, during which the activity more than doubled, owing to the accumulation of the radium emanation. In order the more accurately to compare the two preparations, measurements were taken not only of the two in the same position beneath the electroscope, but also of the stronger preparation at a greater distance, so that the effects compared should be of the same order. The following table shows the actual readings of the electroscope in divisions per minute, corrected for the natural leak (about 4·8), and the ratio between them for the two positions.

Day.	Fraction *A*.	Fraction *AA*.		Ratios.	
1st	10·5	91.5	12·05	8·7	1·14
2nd	13·1	115·0	14·95	8·8	1·14
3rd	15·0	133·0	17·35	8·85	1·15
7th	19·65	170·8	22·3	8·7	1·13
13th	23·9	203·0	26·6	8·5	1·12
34th	26·0	224·0	29·5	8·6	1·14

The fractionation process, from which the fraction *AA* was derived, was continued until twenty-four sets of fractionations had been performed. The products were then combined in two final fractions, the one, labelled *C*, consisting of the weak, and the other, labelled *B*, of the rich fractions. Thus the most active fraction of the original material was obtained in three fractions, *AA*, *B*, and *C*, the weights of which were 2·09, 2·27, and 38 grams respectively, and the relative activities as 4·6:3·85:1. The concentrations of the radioactive matter in the three fractions were therefore as 84:64:1. Fractions *B* and *C* were kept dissolved in water for some days, so that mesothorium-2 should be initially in equilibrium, then evaporated to dryness, sealed up in test tubes, measured immediately, and again after twenty-one days. The relative activity of the two fractions was exactly the same in the two tests, showing that the fractionation process which had altered the concentration of the radioactive matter sixty-two times had not affected the ratio of the two radioactive constituents. For each fraction the proportion of the activity contributed by the mesothorium was almost exactly one-half that contributed by the radium, which is practically the same as that found initially.

This experiment proves conclusively therefore that in the fractional crystallisation of barium chloride, containing mesothorium and radium, the mesothorium and radium behave as a single substance, and there is no hope of separating them by this method.

With the knowledge gained of the chemical nature of mesothorium, a good many further experiments were done on its separation from thorianite, which need not be detailed. They all bore out the view that mesothorium and barium are chemically analogous. It was found that a practically complete separation of the mesothorium and radium from thorium in the thorianite solution could most favourably be effected by adding a small quantity of barium nitrate and a considerable quantity of strong nitric acid, and precipitating the thorium with oxalic acid in the strongly acid solution. The mesothorium is precipitated from the filtrate by pouring it into excess of sodium carbonate solution (which keeps the uranium dissolved), and recovered from the solution of the precipitate in nitric acid by precipitating the barium with sulphuric acid.

In the first experiment with monazite sand, 400 grams were dissolved, by heating it with twice its weight of sulphuric acid and stirring the product with cold water, exactly as in the technical working up of the material. The muddy liquor obtained was decanted from the unattacked sand, which constituted about 20 per cent. of the whole, and left to deposit its sediment. This weighed 4·8 grams, and consisted largely of calcium sulphate. It was labelled *A*. One gram of barium nitrate was dissolved, and added slowly to the clear monazite solution, with efficient stirring. The precipitate (labelled *B*) weighed 1·8 grams. Tested by γ-ray methods, the undissolved sand retained about 4·5 per cent. of the total activity of the material. The β-activities of the sediment *A* and precipitate *B* were as 1 to 3 initially, and as 1 to 2 after forty days. Thus, under the ordinary conditions of the thorium manufacture, an important part of the mesothorium is lost in the insoluble sediments. The chemical behaviour of mesothorium, as is to be expected, is indefinite in the absence of sufficient barium to be quantitatively separable. Monazite contains much less uranium and therefore radium, relatively to thorium, than thorianite, and the γ-rays of the preparation *B* in consequence fell to about half its [sic] maximum value in the course of a month, owing to the effect of the decay of thorium-*X* exceeding the growth of radium-*C*.

Further experiments with 400 and 800 grams of monazite sand were made as before, except that about 1·1 per cent. of barium carbonate was mixed with the sand before heating. The sediment obtained from the muddy solution now contained practically all the mesothorium and radium in the monazite. One such sediment, from 800 grams of monazite sand, weighed 14·5 grams and contained practically the whole of the mesothorium in the material. Its γ-ray activity at the maximum, three days after preparation, was about 70 per cent. of that of the original material. The unattacked sand retained 8 per cent. A further precipitate of 1·6 grams of barium sulphate formed in the clear monazite solution possessed a small initial activity, due to regenerated thorium-*X* only, which almost completely decayed in the course of a month. Certainly less than 5 per cent. of the mesothorium in the sediment was present. As throughout the work thorium-*X*, mesothorium, and radium have always been separated together, the presence of thorium-*X* and absence of mesothorium in this precipitate may be regarded as clear evidence that practically the whole of the mesothorium can be separated from monazite by the method described. The γ-radiation of the main sediment fell to about 57 per cent. of its maximum value after a month, as the effect of the decay of thorium-*X* overpowers the increase due to the generation of radium-*C*.

In the course of two or three years it is to be expected that the preparation will rise in activity to somewhat more than its initial value, due to the regeneration of radiothorium, and in consequence thorium-*X* (and also of radium-*C*, which does not contribute to the initial activity). Then it will decay exponentially, with the period of mesothorium-1, to the constant small proportion contributed by the radium.

A part of this sediment was boiled with sodium carbonate, washed free from sulphates, and dissolved in hydrochloric acid. It left an inactive residue, mainly silica, whilst from the solution practically the whole of the radioactive matter was precipitated with the barium chloride by saturating it with hydrogen chloride. This is further evidence of the resemblance between mesothorium and radium. All the methods effective in the concentration of the latter which were tried serve equally well for mesothorium.

In going over all the measurements, which refer to more than thirty preparations, the activity of which was kept under observation for a month or longer, there is clear evidence also that thorium-*X* is always separated in any chemical operation in the same proportion as mesothorium and radium. It appears that the behaviour observed by Boltwood for the one reaction, precipitation with ammonia, is general. Certain apparent exceptions shown in the preparations measured were found, on referring back to the details of the separation, to be due to a lapse of time after the thorium-*X* had been separated from the mineral before the first measurement. In these circumstances, owing to the decay of the thorium-*X* after separation, its proportionate activity compared with that of mesothorium and radium appears low. Although no separate examination of the point has been made, there is good reason to believe that mesothorium, radium, and thorium-*X* are a chemically inseparable trio. It should be mentioned, however, that G. Hoffmann (*Physikal. Zeitsch.*, 1907, **8**, 553), from a comparison of the coefficient of diffusion and ionic mobility of thorium-*X* in solution, deduced from Nernst's theory that the thorium-*X* ion is singly charged, and is therefore univalent. The conditions under which the radioactive measurements were carried out, however, were very far from definite. In the more recent work of Strömholm and Svedberg (*Zeitsch. anorg. Chem.*, 1909, **61**, 338; **63**,

197), some important additions have been made to the chemistry of the radio-elements. The method employed was new and ingenious. By crystallising various salts in solutions containing radioactive constituents, they sought to determine to which of the known elements the radio-elements were isomorphous. They concluded that no differences existed, even from the quantitative point of view, between thorium-*X*, actinium-*X*, and radium. So far as thorium-*X* is concerned, this agrees perfectly with the results given in this paper. They point out that in the thorium, actinium, and uranium-radium series the three emanations are identical chemically, being members of the family of inert gases. The preceding members, thorium-*X*, actinium-*X*, and radium, are again identical, all being members of the alkaline earth family. Next to these come radiothorium, radioactinium, and ionium, which are all similar, but they are inclined to put in the Periodic Table the respective groups ionium, uranium-*X*, radio-uranium; radioactinium, actinium; and radiothorium, mesothorium, thorium, as analogous to the rare-earth group lanthanum to ytterbium, as follows:

	0.	1.	2.	3–4.	
5th Period ...	Xe	Cs	Ba	La-Yb	
6th Period ...	RaEm	—	Ra	Ionium (UX, RaU)	U
	AcEm	—	AcX	RaAc,Ac	
	ThEm	—	ThX	RaTh,MsTh,Th	

Their work on mesothorium is indefinite and in disagreement, for the most part, with the results in this paper, that mesothorium-1 is identical chemically with radium and thorium-*X*. For example they state that ammonia precipitates all elements of the thorium series except thorium-*X*, and that mesothorium is not precipitated, like radium, along with barium sulphate, citing in support of this last some experiments, which, in their second paper, they withdraw because they have not been able to repeat them. It is clear that the chemical identity of mesothorium with radium completely negatives the above attempt to bring the radio-elements into the Periodic Table. The elements radiothorium, mesothorium, thorium suggest anything rather than the rare-earth group lanthanum to ytterbium.

It appears that chemistry has to consider cases, in direct opposition to the principle of the Periodic Law, of complete chemical identity between elements presumably of different atomic weight, and no doubt some profound general law underlies these new relationships. Apart from the case of the three emanations, for which chemical identity is necessarily a common property of the whole group, we have, in addition to the case of radiolead (210·4) and lead (207·1), which are chemically inseparable, two well-defined groups of triplets: (1) Thorium (232·4), Ionium (230·5), Radiothorium (228·4); (2) Mesothorium-1 (228·4), Radium (226·4), Thorium-*X* (224·4), in which the chemical similarity is apparently perfect. The atomic weights, estimated, for the unknown cases, by subtracting from the atomic weight of the parent substance the known number of helium atoms expelled in their formation, show a regular difference of two units between the successive members of these two groups. The first group consists of quadrivalent elements of the fourth vertical column and the second of bivalent elements of the second column of the Periodic System, and yet the atomic weight of the last member of the first, and the first member of the second, group are, as far as is known, the same.

The chemical identity of the members of the above two groups is almost certainly much closer than anything previously known. In the rare-earth group, elements with neighbouring atomic weights are often so closely allied that they can only be separated after the most laborious fractionation, and distinguished by the difference in their equivalents. But as the latter are always very close, the test is a very rough one in comparison with what is possible for radio-elements. Take, for example, the case of ionium and thorium. Boltwood, Keetman, and, lastly, Auer von Welsbach have all failed completely to concentrate ionium from thorium, the latter after a most exhaustive examination, in which his unrivalled knowledge of the rare-earths was supplemented by the new, powerful methods of radio-active analysis (*Mitteilungen der Radium Kommission*, VI, *Sitzungsber. K. Akad. Wiss. Wien*, 1910, **119**, ii, *a*, 1). The question naturally arises whether some of the common elements may not, in reality, be mixtures of chemically non-separable elements in constant proportions, differing step-wise by whole units in atomic weight. This would certainly account for the lack of regular relationships between the numerical values of the atomic weights.

The examples given include all the known radio-elements with periods of average life longer than a year, except uranium,[15] whilst for this element the fact that it alone gives two α-particles per atom disintegrating, which probably are not derived from two rapidly succeeding changes on account of the lowness of their range, is good ground for considering that uranium may also be a mixture of two chemically non-separable elements in constant proportion due to their genetic relationship, differing in atomic weight by four units. On this view, uranium may be analogous to thorium and radiothorium,

---

[15] Actinium can hardly be considered in this connexion as its chemistry is still relatively imperfectly known.

except that there is no intermediate product of different chemical nature to reveal their separate identities.

It is natural that relationships such as these, even if they were general, should at first appear to be confined to the longer-lived radio-elements. For the short-lived substances, not only on account of the evanescent character of the material is it difficult to determine their true nature. Adsorption plays a much larger part in the separation of the short-lived products than it does in the case of the longer-lived. The reason is not far to seek. Radioactivity is a function, not of mass, but of mass divided by the period of average life. Thus a given amount of an adsorbent may be able to adsorb similar amounts of two radioactive substances before becoming saturated. If, however, the one is much longer lived than the other, when quantities, not equal, but possessing similar radioactivity are acted on, the separation may be practically complete for the shorter-lived substance, and for the other practically inappreciable.[16] Polonium, although its period of average life is less than a year, has well-defined chemical properties, which have been elucidated by the exhaustive investigations of Mme. Curie and Marckwald. It will be interesting to see whether it does not prove to be identical with the still non-isolated "di-tellurium" of Mendeléeff, for the existence of which some recent evidence is forthcoming (W. R. Flint, *Amer. J. Sci.*, 1910, [iv], **30**, 209). It would at least be interesting to apply to the supposed mixtures of tellurium and di-tellurium the methods used by Marckwald in separating polonium from tellurium.

I desire to acknowledge the capable assistance of Mr. W. T. Munro in the preparation of the purified active barium chloride from thorianite.

PHYSICAL CHEMISTRY LABORATORY,
UNIVERSITY OF GLASGOW.

[A theory which covers everything that one knows is promising; if it covers more than one knew originally, it is convincing. Soddy's idea of chemical identity among physically distinguishable, radioactive substances covered excellently the portion of radiochemistry with which he was familiar. Whether it could do more was an interesting question, and

[16] This point of view also explains at once the remarkable observation of Ritzel (*Zeitsch. physikal. Chem.*, 1909, **67**, 725) that a trace of thorium sulphate completely prevents the adsorption of uranium-X by charcoal. For, according to Marckwald and Keetman, uranium-*X* is completely analogous chemically to thorium and cannot be separated from it.

Soddy spent the greater part of 1911 searching the literature to find out. All that he gathered he worked up into a thin book (no more than 92 pages including the index) which was published near the end of the year.

Rutherford and Soddy had insisted when they first proposed the hypothesis of atomic transformation that the radioactive substances were definite elements in the ordinary chemical sense. As such, there should be places for them in the periodic table; indeed attempts had been made to fit them in.[17] Soddy now had firmer knowledge than any of his predecessors, and as he pondered on the proper places for his single-element groups, he began to see regularities in the movement of an atom from group to group as it underwent its successive transformations. This material is set down in the extract from his book which is printed as the next paper.—A. R.]

21

Frederick Soddy

Analogies between the Disintegration Series

[Extracted from *The Chemistry of the Radio-Elements, Part I,* (London: Longmans, Green and Co., 1911) pp. 26–30.]

The disintegration series as they are now known and represented in Fig. 1 [p. 192] offer a glimpse at once as arresting and as elusive as that afforded by the Periodic Table into the unsolved problem of the nature of matter. There is a general analogy between all three series.[18] That between actinium and thorium is almost perfect, except that two products are known between thorium and radiothorium of which no representatives occur in the actinium series. The uranium series is distinguished by three products more at the end unrepresented in either of the other series.

These analogies have become more obvious in consequence of

---

[17] S. Meyer and E. R. von Schweidler, "Untersuchungen über radio-active Substanzen (VI. Mitteilung) Über Radium F (Polonium)," *S. B. Akad. Wiss.* Wien [IIa], 1906, *115*: 63–88.

D. Strömholm and T. Svedberg, "Untersuchungen über die Chemie der radioactiven Grundstoffe II," *Z. anorg. Chem.*, 1909, *63*: 197–206.

A. T. Cameron, "The Position of the Radio-active Elements in the Periodic Table," *Nature, Lond.*, 1909–10, *82*: 67–68.

[18] Hahn and Meitner, Physikal. Zeitsch., 1910, *11*, 493.

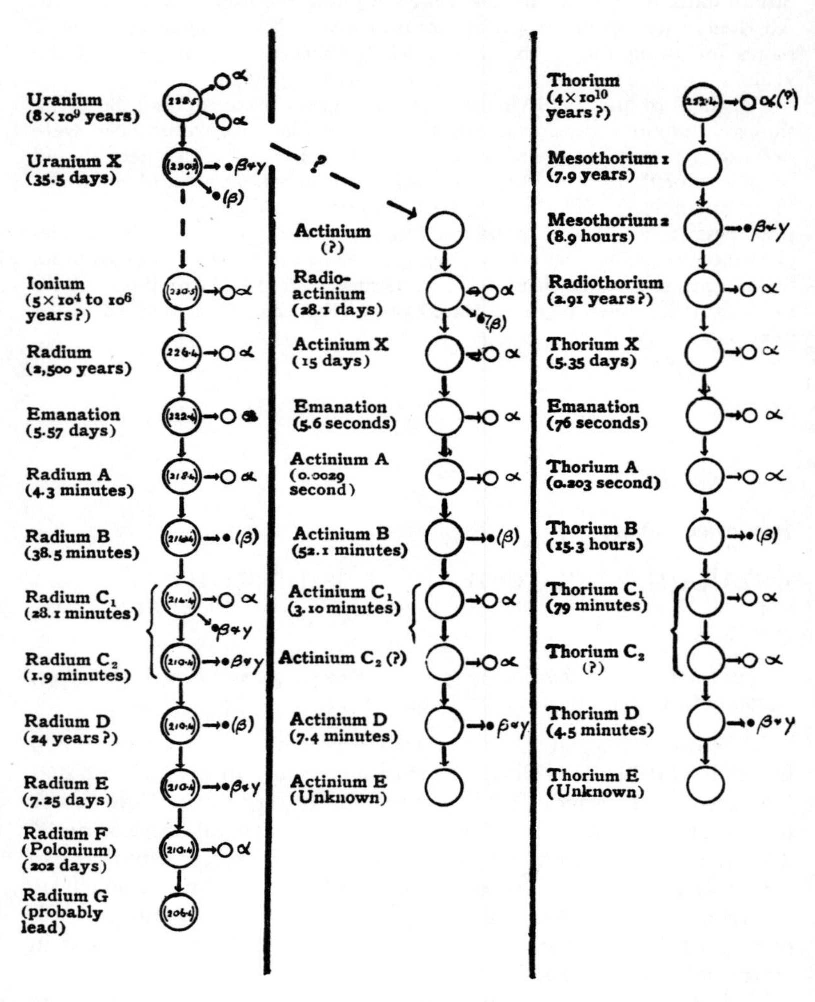

Fig. 1

certain discoveries already alluded to.[19] It has been established that the thorium and actinium emanations both disintegrate with the production of excessively short-lived non-volatile products which give α-rays, but which before have not been separately distinguished from

[19] Rutherford and Geiger, Phil. Mag., 1911 (vi), *22*, 621

the volatile parent. In consequence, the first member of the "active deposits" in each of the three series is now a short-lived α-ray giving product radium A, thorium A, actinium A. The second or B member is a rayless, or more strictly, a (β)-ray giving product, having the longest period of the active deposit group in each case. After this the analogies, though still striking, become less perfect, and it is scarcely profitable, in view of probable new discoveries in this field, to discuss them at length. The next, or C member, in each case gives α-rays, and is somewhat shorter-lived than the preceding member. In all cases it is probably not a single product but consists of two products, indicated by C_1 and C_2. It is doubtful whether these are successive, or simultaneous products. Thorium is the simplest case. The complexity is indicated by two different sets of α-particles being emitted, the range of the C_2 set, 8.6 cm., being the longest known. The product is therefore probably of unmeasurably short period. In actinium indications of the same sort exist, but they are less definite, as the ranges of the particles are nearly alike. In the case of radium only one set of α-rays are emitted, and the range is so high, 6.6 cm., that probably these do not come directly from C_1 but from another very short-lived product. The name, radium C_2, however, has unfortunately been given to a short-lived product giving only β- and γ-rays, and possibly a branch product of C_1. Radium C_1 also, curiously, seems to give the same kind of β- and γ-rays as C_2. In the case of thorium and actinium the typical penetrating β- and γ-rays are given by a subsequent D product of fairly short period. If in the radium series a separate β- and γ-product is ultimately discovered it should be called radium D, and the subsequent products correspondingly renamed. However, radium D is the name given to the long-lived product, giving only feeble (β)-radiation, which obviously is analogous to the stable ultimate E products of the thorium and actinium series, which have not yet been identified. Since radium D is chemically indistinguishable from lead, by analogy the other two products should be stable elements in this part of the Periodic Table.

Disregarding then the minor differences, it may be stated that the A members give α-rays, and are very, sometimes excessively, short-lived, the B members are the longest lived of the active deposit group, and give only unimportant (β)-radiation, the C members give α-rays and are complex, the first representative being rather shorter-lived than the B members. The D members (except radium) are short-lived, and give typical β- and γ-rays. The atoms then re-emerge into the stable or comparatively stable region around lead and bismuth in the Periodic Table. This new nomenclature assists the memory because it makes the names of the analogous members of the "active deposits" correspond in the three cases. Obviously, however, it would be more perfect if the nomenclature had anticipated future discovery by changing the name of radium D to radium E, and of its products correspondingly (October 1911).

After radium D comes a short-lived β-ray product radium E, then the penultimate α-ray giving radium F or polonium, the position of which in the Periodic Table can be fixed from its estimated atomic weight and chemical nature, as occupying the vacant space in the sulphur-selenium-tellurium group of elements, next in atomic weight to bismuth. The ultimate product, radium G, is almost certainly lead. In the thorium and actinium series these members are not represented, the series ending apparently with the E members.

Working now backwards through the series, we have first the three emanations, all of which belong to the zero group of the Periodic Table, and are chemically inert like argon. Next come radium, thorium X, and actinium X, three chemically identical and non-separable elements of the alkaline-earth group. Next come ionium, radio-thorium, and radio-actinium, the latter of which is somewhat indefinite in chemical nature, but the first two of which are identical chemically with thorium of the carbon or tetravalent group and not separable from it. Now another set of suggestive differences in the series appears. Actinium and mesothorium I, the only two really rayless members known, find no analogue in the uranium series, and probably a long-lived rayless product, somewhere intermediate between ionium and uranium, remains to be discovered. Uranium X seems to correspond well with mesothorium 2, both being fairly short-lived and giving penetrating β-rays.

Already, therefore, the disintegration series affords a most remarkable picture of the actual process of the production of the elements from one another, of which the Periodic Law is, as it were, the consequence. Just as from an instantaneous photograph of a waterfall the movement of the apparently motionless water can be inferred, so from the Periodic Law the continuous transformation of the apparently unchangeable elements has been suspected.

Radioactivity has, as it were, cinematographed these transformations, with the result to-day, which none ten years ago could have dared to imagine possible, that in three separate instances we are tracing the successive transit of matter from group to group of the Periodic Table. There are certain points in particular that appear of great significance. The loss of a helium atom or α-particle appears to cause the change of the element, not into the next family but into the next but one. All the products known well enough to classify are of even valency, and this in spite of the fact that the atomic weight of thorium is some six units (or one and a half times the atomic weight of helium) less than that of uranium, and more than that of radium. The families of odd-valency are nowhere represented. Thus we have in each series a well-marked sequence from the tetravalent family (radio-thorium, ionium) into the divalent family (radium, &c.), and into the non-valent family (emanations). Again, the product derived from polonium (group VI.) by the loss of an α-particle is probably not bismuth but lead, in each case the step being from the family of even valency into the next, the family of odd valency being missed. But this is not all. The progress is certainly not so straightforward as this. In several cases the matter appears to *alternate* in its passage, passing through the same family not once but *twice*. The product from thorium (group IV.) is mesothorium (group II.). The product of the latter is radio-thorium (group IV.), which, in turn, produces thorium X (group II.). Again radium D is chemically non-separable from lead (group IV.), its product is polonium (group VI.), while the product of the latter is almost certainly lead (group IV.).

In this connection it should be noted that the helium produced by these changes, which in its ordinary state appears non-valent like argon, carries, in the form of the radiant α-particle, two atomic charges of positive electricity, and is therefore electro-chemically divalent. The passage from group II. (radium) to the zero group (emanation) is direct, whereas the passage from the zero group, the beginning of one period, back to the end of the last period occurs through the long chain of "active deposit" products. In the Periodic Table the transition appears abrupt, and no indication is afforded of intermediate connecting links. The active deposits thus represent a new region in the constitution of matter, of the existence of which there has so far been no evidence. This region bridges the two ends of the Table. The atom having suffered successive reductions of its valency to zero, passes to the electro-negative end of the preceding period through a well-defined sequence of short-lived forms. According to von Lerch's rule, the process is accompanied by a regular increase in the electro-negative character, the successive products being each electro-chemically "nobler" than the last. Obviously the study of the disintegration series is affording fresh light on the Periodic Law, and is showing that new kinds of atoms, nowhere represented by that law, can and do exist, though often they are excessively unstable.

VII. Transformations and Chemical Properties

[Once Soddy had assimilated what was already known in radiochemistry, he wanted to extend that knowledge and enlisted for the work one of his Glasgow students named Alexander Fleck (b. 1889). The first and most obvious question was whether uranium X did really exhibit the chemistry of thorium as Keetman had claimed. By a series of fractionations Fleck found that it did, and then went on to show that radioactinium behaved in the same way and must thus be a fifth variety of thorium.

Following the gaseous emanations in each of the three radioactive transformation series were chains of short-period transformations whose successive products were distinguished by letters of the alphabet. Soddy had suggested that these short-lived substances might belong to "a new region in the constitution of matter, of the existence of which there has so far been no evidence..., [which] bridges the two ends of the Table." Now he realized that they might equally well represent varieties of ordinary elements, even considering their short period of existence. In that case, in an argument by analogy from the transformations which preceded the emanations, the first short-lived products, Ra A, Th A, and Ac A, all formed by alpha-particle emission from the emanations, should be placed in the eighth column of the periodic table. The second group, Ra B, Th B, and Ac B, might then be placed in the sixth column and might even be identical with lead. Of these, thorium B had a half-life of 11 hours, long enough for some quick experiments. Fleck tested thorium B, and found indeed that it was chemically inseparable from lead.

Fleck reported these as preliminary results to the British Association, meeting conveniently in Dundee in that summer of 1912,[1] and returned to his laboratory in Glasgow for further discoveries.

That same fall the Technical Institute in Karlsruhe in Baden admitted a new *Dozent* or Lecturer in physical chemistry and electrochemistry, a 25-year old Pole named Kasimir Fajans. He had been born in Warsaw in Russian Poland in 1887 and had gone abroad for his education, first to Leipzig and then to Heidelberg where he obtained his doctorate in 1909. He had spent a year in Zürich and another in

---

[1] Fleck's paper is summarized in news accounts of the British Association meeting: by an anonymous reporter in *Chem. News*, 1912, *106*: 128; and by G. von Hevesy (then at Manchester) in *Z. Elektrochem.*, 1912, *18*: 865.

Rutherford's laboratory in Manchester, where he had been kept busy in the tangle of short-lived products which followed the emanations. From these investigations he had drawn his Inaugural Dissertation, the long, scholarly work by which he demonstrated his competence to lecture.[2]

With that task finished, he set himself a new goal: to discover the order which underlay the chains of the radioactive transformations. As a starting point he had the material in Soddy's book (which he had reviewed for *Le Radium*,[3] the French journal edited by Pierre Curie's former assistant, Jacques Danne) and a general principle, rather too simple to be true, which was cited sometimes under the name of Lucas and sometimes under that of von Lerch.

The elements along the left edge of the periodic table enter aqueous solution as positive ions and react vigorously with oxygen. Those along the right edge go in as negative ions and oxidize far less readily. Within, the extremes shade into one another; an eye travelling across a row of the periodic table from left to right will encounter elements which are progressively less electropositive and reactive with oxygen, hence progressively more electronegative or nobler. The rule of Lucas or von Lerch stated that in every radioactive transformation the new element produced was more electronegative, or nobler, than the original element. As Fajans perceived, this principle was true for transformations in which a beta particle was emitted. For alpha-particle transformations, the new element was more electropositive.

In the first of the two papers which follow, Fajans tested his two principles critically for every transformation about which he could collect information. In the second, he used them as guides for placing all the radioactive elements in the periodic table. Except at one awkward point, it can be seen that they worked admirably. Fajans's rules call for a monotonic relation between ray emission and change in electrochemical nature, but as it happens, the elements of the natural radioactive series occupy positions at the ends of two rows of the periodic table. In the transition from row to row, there must be accommodation for a periodic electrochemical change. To avoid this accommodation, Fajans found it necessary to postulate the existence of three first-column alkali metals (Ra X, Th X_2, Ac X_2) to stand between the electropositive, alkaline-earth isotopes of radium (Ra, Th X, Ac X) and the noble-gas emanations which he chose to consider as extremely electronegative. With this exception, however, he succeeded admirably in assigning the elements to the places they still occupy.

It may be worth pointing out to the modern reader that Fajans was working here entirely from chemical evidence. He had no comprehensive theory of atomic structure to guide him. At the time when these two papers were printed, on February 15, 1913, Rutherford had indeed published a theory of alpha-particle scattering based on a nuclear

---

[2] K. Fajans, "Die Verzweigung der Radiumzerfallsreihe," *Verhandlungen des naturhistorisch-medizinschen Vereins zu Heidelberg* [N. F. XII], 1912, *2*: 173–241.

[3] K. Fajans, *Radium, Paris*, 1912, *9*: 238–239.

model of the atom,[4] but no experimental test of that theory had yet appeared.[5] Niels Bohr in Copenhagen was actually working on the problem of a one-electron, nuclear atom, but it would be six weeks more before his paper was finished and five months before it would be in print. Bohr's second paper dealing with more complicated atoms lay still farther in the future and the idea that nuclear charge might be the same as atomic number would form one of its hypotheses. When that second paper appeared the following September it would quote the rules which Fajans gives here, presenting them in fact as persuasive evidence for the truth of its special hypothesis about nuclear charge.[7]—A. R.]

22

K. Fajans

On a Relation between the Nature of a Radioactive Transformation and the Electrochemical Behavior of the Radioelement Involved

[Translation of "Über eine Beziehung zwischen der Art einer radioaktiven Umwandlung und dem elektrochemischen Verhalten der betreffenden Radioelemente," *Physikalische Zeitschrift*, 1913, *14*: 131–136 (15 February).]

The radioelements, which are accessible for the most part only in extraordinarily small quantities, have their electrochemical properties defined by their tendency to be deposited on ordinary metals dipped into their solutions, or by their behavior in electrolysis. To draw conclusions about the normal potential from the first of these experiments, as has frequently been undertaken, and in this way to make comparison with other elements possible, one must not only bring into

---

[4] E. Rutherford, "The Scattering of α and β Particles by Matter and the Structure of the Atom," *Phil. Mag.* [6], 1911, *21*: 669–688.

[5] H. Geiger and E. Marsden, "The Laws of Deflexion of α Particles Through Large Angles," *Phil. Mag.* [6], 1913, *25*: 604–623 (April issue).

[6] N. Bohr, "On the Constitution of Atoms and Molecules," *Phil. Mag.* [6], 1913, *26*: 1–25 (July issue).

[7] N. Bohr, "On the Constitution of Atoms and Molecules. Part II—Systems Containing only a Single Nucleus," *Phil. Mag.* [6], 1913, *26*: 476–502 (September issue).

the calculation the concentration of the ions being considered (the radioactive as well as those of the metal dipped in[8]), but one must also be oriented to the extraordinarily small quantity of the radioactive element which deposits on the metal. A coherent layer of metal for example, shows an electrochemical potential quite different from that of a dilute solid solution of one metal in another. Concerning this state we are still not clear.

However, the experience obtained up to now in this field is completely satisfactory for a qualitative comparison of the electrochemical behavior of two radioelements present in not very different concentrations. One surely does not go wrong to conclude with F. von Lerch[9] that RaC is electrochemically nobler than RaB, given the fact that from a weakly acid solution of *RaB* + *RaC*, the latter can be obtained on a sheet of nickel practically free of the former. From the observations of E. Meyer and E. von Schweidler that polonium (*RaF*) is deposited on the cathode at smaller current densities than *RaE*, and that *RaD* requires greater current densities than these two elements, one can conclude also that *RaE* is more electronegative than *RaD* and that *RaF* is the noblest of these three metals. Finally one can draw conclusions about the electrochemical behavior of many radioelements from their complete chemical equivalence with certain ordinary elements. So, for example, from the impossibility of separating ionium from thorium,[10] there follows their complete chemical and hence electrochemical analogy. Thanks to these methods it is now possible to compare the electrochemical properties of almost every radioelement.

As a result of such a comparison, R. Lucas[11] pronounced a rule some time ago that each radioactive transformation leads to a product which is electrochemically nobler than its immediate parent. In a radioactive series, then, the electronegative character must continually increase from the beginning to the end of the series.

On the basis of more recent measurement, G. von Hevesy[12] and the author[13] showed simultaneously and independently that this is not always the case.

Recently it was announced by the author[14] that there is another

---

[8] In this respect the beautiful work of G. von Hevesy, Phil. Mag. **23**, 628, 1912, signifies an important advance.

[9] Ann d. Phys. (4) **20**, 345, 1906.

[10] Fr. Soddy, Die Chemie der Radioelemente, 1912, p. 81.

[11] This journal **7**, 340, 1906.

[12] This journal **13**, 672, 1912.

[13] Le Radium **9**, 239, 1913.

[14] K. Fajans, Appendix to the Inaugural Dissertation, Karlsruhe, 1912. Also, Verhandl. d. Naturhist. Mediz. Vereins zu Heidelberg **12**, p. 235.

regularity which is valid in every case investigated. In radioactive transformations *which are linked with the emission of beta rays, a product is formed which is more electronegative than the direct parent; on the other hand, exactly the opposite proves correct for transformations which proceed with the emission of alpha rays, in such cases the immediate parent is electrochemically nobler than the transformation product.*

It is the aim of this paper to show the correctness of these rules on the basis of the existing experimental material.

The radioactive transformations known today are displayed in the following diagram, taking into account some conclusions drawn in this and the following work.[15] For the products of the active deposits of thorium and actinium the new nomenclature has been used.[16]

The numbers under the name of each element signify its half-value time, those above indicate the group in the periodic system to which the element in question belongs, in the few cases where it is known. Particulars on the last point are to be found in the book by Soddy, which is cited. The letters α and β above the arrows indicate with which ray the transformation is connected. The gamma rays, which play no role in our problem, are not indicated.

As the series referred to show, there are only three transformations which seem to be connected with both kinds of rays; these are radium → radium emanation, thorium X → thorium emanation, and radioactinium → actinium X. It has turned out in many other cases however that a complex radiation which at first was ascribed to a simple transformation, actually belonged to several transformations. It is therefore very probable[17] that in these cases also the processes concerned are not simple; it may be a question of two successive transformations or a double transformation of the same product (branching of the series[18]). These cases will thus be considered separately.

There are also two transformations for which up to now no ray has been detected: mesothorium I → mesothorium II and actinium → radioactinium. By analogy with other such cases[19] it can be assumed that in these transformations very soft beta rays, as yet undetected, appear. The hypothetical character of these rays is indicated in our scheme by parentheses.

---

[15] In the branching of the thorium series, for simplicity the scheme of Marsden and Darwin is indicated. It would change nothing however in the working out of our ideas if we took the more correct scheme of Miss Meitner. For particulars, see the Inaugural Dissertation of the author.

[16] E. Rutherford and H. Geiger, Phil. Mag. **22**, 621, 1911.

[17] Cf. O. Hahn and L. Meitner, this journal **11**, 497, 1910.

[18] Cf. the Inaugural Dissertation of the author.

[19] Cf. O. von Baeyer, O. Hahn, and L. Meitner, this journal **12**, 378, 1911.

$$\underset{5\times 10^{9}\text{ y.}}{\overset{6}{UrI}} \xrightarrow{\alpha} \underset{24.6\text{ d.}}{\overset{4}{UrX}} \xrightarrow{\beta} \underset{(10^{6}\text{ y.})}{\overset{6}{UrII}} \xrightarrow{\alpha} \Big| \underset{(10^{6}\text{ y.})}{\overset{4}{Io}} \xrightarrow{\alpha} \underset{2000\text{ y.}}{\overset{2}{Ra}} \xrightarrow{\alpha,\beta} \underset{3.86\text{ d.}}{\overset{0}{Rad\,Em}} \xrightarrow{\alpha} \underset{3\text{ m.}}{RaA} \xrightarrow{\alpha} \underset{26.7\text{ m.}}{RaB} \xrightarrow{\beta} \underset{19.5\text{ m.}}{RaC_1}$$

$$RaC_1 \xrightarrow{\alpha} \underset{1.4\text{ m.}}{RaC_2} \xrightarrow{\beta} ?$$

$$RaC_1 \xrightarrow{\beta} \underset{(10^{-6}\text{ s.})}{RaC'} \xrightarrow{\alpha} \underset{16\text{ y.}}{\overset{4}{RaD}} \xrightarrow{\beta} \underset{5\text{ d.}}{RaE} \xrightarrow{\beta} \underset{134\text{ d.}}{RaF} \xrightarrow{\alpha} \text{Lead}$$

$$\underset{3\times 10^{10}\text{ y.}}{\overset{4}{Th}} \xrightarrow{\alpha} \underset{5.5\text{ y.}}{\overset{2}{MesThI}} \xrightarrow{(\beta)} \underset{6.2\text{ h.}}{MesThII} \xrightarrow{\beta} \Big| \underset{2\text{ y.}}{\overset{4}{RadTh}} \xrightarrow{\alpha} \underset{3.7\text{ d.}}{\overset{2}{ThX}} \xrightarrow{\alpha,\beta} \underset{53\text{ s.}}{\overset{0}{ThEm}} \xrightarrow{\alpha} \underset{0.14\text{ s.}}{ThA} \xrightarrow{\alpha} \underset{10.6\text{ h.}}{ThB} \xrightarrow{\beta} \underset{55\text{ m.}}{ThC_1}$$

$$ThC_1 \xrightarrow{\alpha} \underset{3.1\text{ m.}}{ThD} \xrightarrow{\beta} ?$$

$$ThC_1 \xrightarrow{\beta} \underset{(10^{-11}\text{s.})}{ThC_2} \xrightarrow{\alpha} (ThD_2)$$

$$Act \xrightarrow{(\beta)} \Big| \underset{19.5\text{ d.}}{RadAct} \xrightarrow{\alpha,\beta} \underset{10.2\text{ d.}}{\overset{2}{ActX}} \xrightarrow{\alpha} \underset{3.9\text{ s.}}{\overset{0}{ActEm}} \xrightarrow{\alpha} \underset{0.002\text{ s.}}{ActA} \xrightarrow{\alpha} \underset{36.1\text{ m.}}{ActB} \xrightarrow{\beta} \underset{2.15\text{ m.}}{ActB} \xrightarrow{\alpha} \underset{4.7\text{ m.}}{ActD} \xrightarrow{\beta} ?$$

$$\overset{1}{\mathrm{K}} \xrightarrow{\beta} ? \qquad \overset{1}{Rb} \xrightarrow{\beta} ?$$

Our task is made easier in certain cases because it is possible to deduce without hesitation from the behavior of the terms of one series the behavior of the terms in the other series which correspond from the radioactive point of view (genetic relationship, approximate life, character of the rays), but for which the electrochemical properties cannot be directly determined for any reason. From ionium, radiothorium and radioactinium onward, the terms of the three series which correspond in radioactivity are arranged in the same vertical column, and in all the cases studied these products correspond completely from the chemical and electrochemical point of view.[20]

After these introductory remarks it will now be shown that in all the cases where the nature of the transformation is known with certainty and where the electrochemical behavior of the two elements involved has been closely studied, the two rules mentioned at the beginning hold without exception.[21] We shall begin with beta-ray transformations, and it will be shown that here the transformation product is always more electronegative than the parent substance. For the following transformations they appear directly from the works cited.

$$RaB \xrightarrow{\beta} RaC_1 \ (5, 6, 7) \qquad ThB \xrightarrow{\beta} ThC_1 \ (2, 6)$$

$$ActB \xrightarrow{\beta} ActC \ (4, 6)^{22} \qquad RaD \xrightarrow{\beta} RaE \ (3, 7)$$

$$RaE \xrightarrow{\beta} RaF \ (3).$$

Concerning the transformation mesothorium I $\xrightarrow{\beta}$ mesothorium II $\xrightarrow{\beta}$ radiothorium, the following may be said. Mesothorium I behaves entirely like radium (1 p. 133), therefore like a very weakly noble, alkaline earth metal. Mesothorium II must be equivalent to

---

[20] Cf. F. Soddy, Die Chemie der Radioelemente, 1912, p. 55; G. von Hevesy, Phil. Mag. **23**, 628, 1912.

[21] Here are listed by number the works often cited:

1. F. Soddy, Die Chemie der Radioelemente, 1912.
2. F. von Lerch, Jahrb. d. Radioakt. u. Elektronik, **2**, 470, 1905.
3. E. Meyer and E. von Schweidler, Wien. Ber. **115**, (IIa), 698, 1906.
4. M. Levin, this journal **7**, 812, 1906.
5. F. von Lerch, Ann d. Phys. (4) **20**, 345, 1906.
6. G. von Hevesy, Phil. Mag. **23**, 628, 1912.
7. K. Fajans, Inaugural Dissertation, Karlsruhe, 1912; also, Verhandl. d. Naturhist. Mediz. Ver. zu Heidelberg **12**, 235.
8. G. von Hevesy, this journal **13**, 672, 1912.
9. F. von Lerch and von Wartenberg, Wien. Ber. **118** (IIa), 1575, 1909. [Sic, should be von Wartburg—A. R.]

[22] To prevent misunderstandings, it must be mentioned that in the older works (2 and 4) the present *B*- and *C*-products of the thorium and actinium series bear the names of *A* and *B*.

the rare earths in its behavior (1 p. 136), while radiothorium is completely analogous to thorium (1 p. 138) and thus belongs in the fourth group of the periodic system. Therefore mesothorium II like radiothorium is nobler than mesothorium I. In the transformation of mesothorium I to mesothorium II we have assumed the existence on other grounds of very soft beta rays. Thus it stands in good agreement that mesothorium II is nobler than mesothorium I. A direct comparison of mesothorium II with radiothorium in respect to their electrochemical behavior cannot be accomplished with certainty with the existing material; perhaps it can be concluded from the following that as our rule demands, radiothorium is nobler than mesothorium II. M. Levin (4) found that radioactinium, which corresponds to radiothorium, could be deposited electrolytically from weakly acid solution, while Miss Meitner[23] states that mesothorium II is deposited only in completely neutral solution. A conclusive decision can be obtained only by a quantitative investigation. The transformation actinium $\xrightarrow{(\beta)}$ radioactinium should be linked with the emission of soft beta rays. In its behavior, actinium stands between lanthanum and calcium (1 p. 161); it is thus less noble than radioactinium, which corresponds radioactively as well as chemically to radiothorium in the fourth group of the periodic system (1 p. 163). Thus here also agreement with the rule is to be noted.

It is also not without interest to remark that the transformations of the alkali metals *Rb* and *K* can only lead to elements more noble than themselves. Here too we have beta-ray transformations as the rule requires.

In the three cases $RaC_2 \xrightarrow{\beta}$?, $ThD \xrightarrow{\beta}$?, $ActD \xrightarrow{\beta}$? the rule cannot be tested since the transformation products are not known. On the other hand, it could be predicted on the basis of these rules that these products will be nobler than the parent substances.

Before we pass on to the other beta-ray transformations, the alpha-ray transformations should be discussed. For these it will now be shown that *the transformation leads to a product which is more electropositive than the parent.*

For the transformation thorium $\xrightarrow{\alpha}$ mesothorium I, the facts show directly that the first belongs to the fourth group of the periodic system while mesothorium I, as already mentioned, is an alkaline earth metal. Ionium (1 p. 81) and radiothorium are completely like thorium, while radium and thorium X (1 p. 141) are alkaline earth metals, whence it is shown that the rule is valid also for the transformations $Io \xrightarrow{\alpha} Ra$ and radiothorium $\xrightarrow{\alpha}$ ThX. For the transformations

---

[23] This journal **12**, 1094, 1911.

$ThA \xrightarrow{\alpha} ThB$ and $ActA \xrightarrow{\alpha} ActB$ the rule cannot be tested directly since it is impossible to study the electrochemical behavior of elements as short-lived as *ThA* and *ActA*. It may certainly be assumed, however, that the relations here correspond entirely with those in the analogous case $RaA \xrightarrow{\alpha} RaB$. RaB is less noble than *RaA* (7, 8), hence the validity of our rule is shown for these three cases. For the three corresponding (cf. 7) cases $RaC_1 \xrightarrow{\alpha} RaC_2$, $ThC_1 \xrightarrow{\alpha} ThD$, and $ActC \xrightarrow{\alpha} ActD$, only the second has been studied electrochemically (9) and the experiments show that *ThD* is less noble than ThC_1 as the rule requires. This may be assumed also for the two other cases. The alpha-ray transformation of polonium (*RaF*) leads most probably to lead. Now *RaD* is entirely similar to lead (1 p. 115). *RaF*, which we have seen is much nobler than *RaD*, is therefore nobler than its alpha-ray transformation product, lead.

The alpha-ray transformations of the three emanations into the corresponding A-products are still to be discussed. The emanations are chemically inert, noble gases; thus their electrochemical behavior cannot be determined by ordinary methods. However if electrochemical character is defined as the inclination of the atom to split off or take on negative electrons, it may properly be said that the emanations are much more electronegative than any of the other radioelements, which are all metals. Otherwise they would be qualified to form cations. That such a conclusion is perhaps justified is shown by the following: from the fact that the noble gases are not able to form anions, one must conclude in the same way that they have no tendency to add electrons. It is very much worth remarking that J. Franck[24] was able to conclude from mobility measurements that in noble gases free electrons are considerably more stable than in any other gases, which therefore have a much greater inclination to add electrons than the noble gases. This tendency to add on is especially great for chlorine, which is known from chemistry to be strongly electronegative.[25]

On the basis of this wider conception of electrochemical character, the emanations are more electronegative than the remaining radioelements. Now the transformations of the emanations into the A-products are accompanied by alpha rays and our rule is therefore applicable to these three cases.

It has now been shown that in all the cases where both the nature of the transformation and the electrochemical behavior of the two elements are well known, our two rules hold without exception.

---

[24] Verh. d. D. Phys. Ges. **12,** 613, 1910.

[25] Cf. J. Franck and P. Pringsheim, Verh. d. D. Phys. Ges. **13**, 328, 1911.

Thus it will not be too bold to use these rules to explain other, not yet sufficiently studied transformations.

The transformations

$$UrI \xrightarrow{\alpha} UrX \xrightarrow{\beta} UrII \xrightarrow{\alpha} Io$$

will now be discussed. The existence of *UrII* follows only from the fact that uranium emits two alpha particles of different ranges.[26] Its chemical nature appears to be very close to that of uranium (I); it belongs therefore, like it, to the sixth group of the periodic system. On the other hand *UrX* and ionium show a very wide-ranging chemical correspondence and are both to be placed beside the tetravalent thorium. Hence it follows that *UrX* and *Io* are much less noble than *UrI* and *UrII.* The genetic connections between these four elements are not yet entirely clear. There are two possibilities. The corresponding transformations might correspond to the scheme

$$UrI \xrightarrow{\alpha} UrII \xrightarrow{\alpha} UrX \xrightarrow{\beta} Io.$$

This would contradict our rule that *UrX* is less noble than *UrII.* However the fact that *UrI* and *UrII* on one side and *UrX* and *Io* on the other can be so little distinguished electrochemically, makes this scheme improbable. In all other cases the jump in electrochemical behavior in either an alpha-ray or a beta-ray transformation is considerable. The second possible scheme is that given above, and this, as Dr. von Hevesy has told me, is also probable on other grounds. This scheme corresponds at every point to our two rules and should be assumed to be correct. Also only this scheme can be brought into harmony with the methods of the following paper.

It is worth remarking on the consequences which our rules yield for the transformations radium $\xrightarrow{\alpha,\beta}$ radium emanation and thorium X $\xrightarrow{\alpha,\beta}$ thorium emanation, of course only on the assumption that electrochemical considerations with respect to the noble gases are inadmissible. As was mentioned at the beginning of this paper, Hahn and Meitner have concluded from the complex radiation that the transformations involved are complex. Our two rules now permit us to deduce that monovalent alkali metals should exist between the divalent alkaline earth metals radium and *ThX* on one side and the non-valent emanations on the other. The disintegration of radium and thorium X must be connected with the emission of the alpha rays and would lead to the less noble alkali metals, while the disintegration of the latter represent beta-ray transformations and in agreement with the rule lead to the much nobler emanations.

---

[26] Geiger and Nuttall, Phil Mag. **22**, 439, 1912.

As concerns the transformation actinium X $\xrightarrow{\alpha}$ actinium emanation, it must be concluded by analogy that here also beta rays must be present which because of their softness have so far remained undiscovered. Here is the only apparent contradiction of our rule.

The case of radioactinium $\xrightarrow{\alpha,\beta}$ actinium X, where also both alpha and beta rays seem to appear in one transformation, will be treated briefly in the following paper.

There remain to be discussed now only the transformations $RaC_1 \xrightarrow{\beta} RaC' \xrightarrow{\alpha} RaD$ and $ThC_1 \xrightarrow{\beta} ThC_2 \xrightarrow{\alpha}$?, in which the electrochemical character of RaC' and ThC_2 cannot be determined directly on account of their extraordinarily short lives. From our rule it can be deduced that RaC' is nobler than RaC_1 and also RaD. The same applies to ThC_2.

The regularities found now demonstrate clearly that the same characteristics in the structure of the atom which condition the nature of a radioactive transformation also find expression in the tenacity of binding of the electrons which are responsible for chemical processes. So simple a connection was by no means to be expected, for everything we know up to now about the radioactive process would entitle us to assume that in it an entirely different region of the atom comes into consideration than in chemical processes. A few attempts which have been made to interpret these regularities have not yet led to satisfactory results.

The consequences which follow from the new rules for the placing of the radio-elements in the periodic system are dealt with in the following paper.

Summary

It is shown that the following two rules possess validity in all cases where they can be tested directly:

1. In an alpha-ray transformation, the resulting product is electrochemically more positive than the direct parent substance.
2. In a beta-ray transformation the transformation product is electrochemically more negative than its parent substance.

(Received 31 December 1912.)

23

K. Fajans

The Placing of the Radioelements in the Periodic System

[Translation of "Die Stellung der Radioelemente im periodischen System," *Physikalische Zeitschrift*, 1913, *14*: 136–142 (15 February).]

In the three known radioactive transformation series, those products which possess concordant properties also correspond completely from the chemical point of view. This means that the sequence in which the radioactive transformations run through the groups (vertical columns) of the periodic system is the same in all three cases (with the exception of the first terms). No one has yet been able to decide what groups these are for every transformation, and the problem of a final arrangement of all the radioelements in the general periodic table is thus unsolved. In part the difficulties lie in the following. It would be expected that the transformations which lead from elements of higher atomic weight to those of lower would simply pass across all the groups of the periodic system from right to left. However this is not correct. According to Soddy,[27] who has recently discussed this question, only elements of the even groups (6, 4, 2, 0) are represented in the radioactive series, which must mean that the transformations always leap over the odd groups. Also, the direction of progress through the various groups does not remain always the same; it goes, for example (see the table in the preceding paper) from the fourth group (thorium) to the second (mesothorium I), then back to the fourth (radiothorium), and once again back to the second (thorium X). The first of these two rules, as we shall see directly, is not correct in general. The second is accurate without a doubt. The two rules demonstrated in the preceding work concerning the electrochemical behavior of successive radioelements now permit us to find an explanation of these oscillations between the groups of the periodic system.

For alpha-ray transformations, the rule applies that a product is produced which is more electropositive than its direct parent. Now the electropositive character increases in the horizontal row of the periodic table from right to left, from the higher to the lower groups.

[27] Die Chemie der Radioelemente, 1912, p. 61.

It is thus possible to formulate this rule: *that by an alpha-ray transformation a product is formed that belongs in a lower group of the periodic table than its parent.* In the table of the preceding paper will be found the proof that this is indeed correct in every case of an alpha-ray transformation where the chemical character of the elements involved is known. This passage from a higher to a lower group corresponds absolutely with the fact that in an alpha-ray transformation the atomic weight is diminished (by the atomic weight of helium).

The question still arises by how many groups to the left this displacement occurs. In the cases studied, there is observed, as Soddy (*loc. cit.*) first emphasized, and as can be seen from the table of the preceding paper, a jump for alpha-ray transformations to the second adjacent column in the same horizontal row, and it will now be assumed that this is correct for all alpha-ray transformations.[28]

As concerns the beta-ray transformations, it must be assumed that they deal only with a rearrangement of the constituent parts of the atom while the atomic weight remains unchanged. The periodic system fails entirely to help if we wish to derive from it the direction in which the passage for these transformations occurs. It does not provide in any way for elements of the same atomic weight which differ in chemical behavior. That in beta-ray transformations the parent differs chemically from the transformation product is shown in the cases mentioned in the preceding paper, and is evident also from the following. The decision as to the direction in the periodic system in which the change proceeds follows from the rule derived for beta-ray transformations, that they lead to a product which is more electronegative than the parent; *that means that in a beta-ray transformation a transition into a higher group of the periodic system must occur, that is toward the right*, since the electronegative character of the elements increases from left to right. The few cases in which this conclusion can be tested show complete agreement. In the case

$$RaD \xrightarrow{\beta} RaE \xrightarrow{\beta} RaF$$

there is produced by two beta-ray transformations from *RaD*, which corresponds entirely to lead and thus belongs in the fourth group, *RaF*, which is to be compared with tellurium in the sixth group. With the transformations

$$Mes\,ThI \xrightarrow{\beta} Mes\,ThII \xrightarrow{\beta} \text{Radiothorium}$$

[28] Only the transformation

$$\text{radium} \xrightarrow{\alpha} \text{(alkali metal)} \xrightarrow{\beta} \text{emanation}$$

and the corresponding transformations of the other series offer difficulties, where by the assumptions of the preceding work a crossing takes place from divalent radium to a monovalent alkali metal. This point will be discussed elsewhere.

from *MesThI* (second group) radiothorium in the fourth group is formed by two beta-ray transformations. From actinium, which stands by its behavior between calcium and lanthanum and thus belongs to the second or third group, radioactinium is formed by a beta-ray transformation; on the grounds of likeness with radiothorium it is to be assigned a place in the fourth group. As with the alpha-ray transformations, the question arises here as to how many groups to the right in the periodic system the displacement for a beta-ray transformation covers. In the two cases first mentioned a displacement over two groups occurs for two successive beta-ray transformations. The simplest assumption is of course that such a transformation produces a crossing into the next group. We shall assume this rule also as generally valid. On the basis of these two rules, according to which *in every alpha-ray transformation a displacement occurs over two groups from the right to left in the periodic system, and in every beta-ray transformation, over one group from left to right in the same horizontal row,* we can now attempt to assign to the elements not yet closely characterized their places in the periodic system.

From what has just been said, it follows that mesothorium II must belong to the third group; which is in complete harmony with the statement that it behaves like an element of the group of the rare earths (1 p. 136). To radium E from the foregoing must be assigned a place in the fifth group. A serious question now arises concerning the places in the periodic system which belong to the short-lived products of the active deposits of the three series. The investigation of their chemical nature runs into difficulties because of their short lives. Soddy (1 p. 62) has expressed an original view on this point. In the periodic system, radium emanation is at the left end of a horizontal row, while polonium, which belongs in the sixth group, is at the right end of the next horizontal row above. The products of the active deposit, according to Soddy, form a connecting series between these two rows for which no analog exists among the ordinary elements. An assumption of such a new type of element is not necessary, however, as will be shown directly. That is to say if one follows the radium series backward from *RaD* which belongs in the fourth group, and applies the two rules just proposed, one finds for the elements in the active deposit of radium, and thus for the corresponding products of the two other series, the following places: *RaD*, fourth group, *RaC′* (ThC_2) sixth, RaC_1 (ThC_1, *ActC*) fifth, *RaB* (*ThB*, *ActB*) fourth, and *RaA* (*ThA*, *ActA*) again the sixth group of the periodic system. The transformation of the emanation leads thus from the zero group to the sixth in the next horizontal row. If the eighth group, which occupies a special position, is not considered, then this alpha-ray transformation also jumps over one group (the seventh). These conclusions can be tested by experience only in the case of the B-products: the requirement that they must belong in the fourth group stands in complete agreement with the statement of A. Fleck (Zeitschr. f. Elektrochem. *18*, 865, 1912) that thorium B is inseparable from lead, and also with the fact that *RaB*[29] is equivalent to *RaD*[30] of the fourth group in that it is precipitated with barium sulfate. It can even be separated from radium C_1 in this way.

If now one writes down the groups to which, according to these ideas, the products from *RaA* to lead belong, one gets the series: 6, 4, 5, 6, 4, 5, 6, 4. The grouping 6, 4, 5, 6, 4 then repeats itself twice. It is decidedly worth remarking that a similar analysis of the beginning of the uranium series

$$\underset{6}{UrI} \rightarrow \underset{4}{UrX} \rightarrow \underset{(5)}{?} \rightarrow \underset{6}{UrII} \rightarrow \underset{4}{Io}$$

also produces the series 6, 4, (5), 6, 4, if one considers *UrX* as complex and made up of two successive beta-ray emitting products. Such an assumption is thoroughly justified since *UrX* possesses two groups of beta rays, one very hard, such as is found with very short-lived elements, the other very soft, which corresponds well with the long-lived *UrX* (half-value time 24.6 days). This hypothetical, short-lived element must belong in the fifth group of the periodic system and stand in the transformation series between *UrX* and *UrII.* It should be emphasized that only the foregoing scheme, and not the scheme

$$\underset{6}{UrI} \xrightarrow{\alpha} \underset{6}{UrII} \xrightarrow{\alpha} \underset{4}{UrX} \xrightarrow{\beta} \underset{4}{Io}$$

can be reconciled with the rules for the changing of groups in alpha- and beta-ray transformations. From the foregoing it follows directly that in a radioactive series the nature of the transformations also changes periodically.

There remain now only the corresponding elements RaC_2, *ThD*, and *ActD* to be assigned to places in the periodic table. They are formed from RaC_1, ThC_1, and $Act\,C_1$, which from the foregoing, belong in the fifth group. After an alpha-ray transformation, according to Soddy's rule as here generalized, they ought to belong to the third group. The strong electropositive behavior of *ThD* is in excellent agreement with that. Concerning *ActD*[31] nothing can be affirmed since we are not sufficiently oriented toward either its chemical nature or its genetic relationships.

---

[29] F. v. Lerch, Ann d. Phys. (4) **20**, 345, 1906.

[30] G. N. Antonoff, Phil. Mag. **19**, 825, 1910.

[31] [The original has *UrJ*, which is nonsense in this context and is probably a compositor's misreading of a handwritten *AçD*.—A. R.]

After it has been found in which groups of the periodic system all the radioactive products belong, an attempt will be made to clarify their relationship to the system in general. In the adjacent table [p. 212] all the radioelements are arranged in the order of decreasing atomic weights in the groups to which they belong. The atomic weight of uranium, 238.5, and of thorium, 232.4, serve as bases for the calculation of the atomic weights, which were calculated on the assumption that in an alpha-ray transformation the atomic weight is lowered by 4.0, while in a beta-ray transformation no change in the atomic weight occurs. The atomic weight of actinium is still unknown. This question will be discussed at the close of the paper.

The most striking fact that this table shows is that places in the periodic system that ordinarily belong to only one element are here occupied by several, in many groups up to six. However all the elements of one group (the difference in their atomic weights always comes to 2 units) are so thoroughly like one another that they cannot be separated from one another by chemical means or even by crystallization. That holds for all cases in which attempts at separation have been undertaken, for example with *UrI* and *UrII*, ionium and thorium, mesothorium I and radium, etc. A mixture of such elements, which for example are always extracted from the minerals together, behaves chemically like a single element. There are now two possible points of view for arranging such a complex element in the periodic system. One might assign it an atomic weight which would be a mean of the atomic weights of its constituent elements. The following way, however, is more natural. If one isolates such a complex element and determines its atomic weight experimentally, one would of course find a value which lay closest to the atomic weight of whatever simple element was present in the greatest amount. That, however, would be the longest-lived element, since the longer-lived an element is, the greater the quantity in which it is present in radioactive equilibrium. One can then, in those cases in which the lives of the elements in a group are very different, insert the atomic weight of the longest-lived. If one proceeds according to these two methods, one obtains the following arrangements [p. 213], the first of which indicates the mean of the atomic weights, while the second considers only the longest-lived element of each group. The actinium series is not considered here.

One notices at once that these two arrangements have this in common with the periodic law for the ordinary elements: the *atomic weights of the elements in the same horizontal row decrease regularly from the higher groups to the lower*, in contrast to the radioactive transformation series in which, as we have seen, with falling atomic weight an oscillation is to be observed between the different groups. This arrangement

0	I		II		III		IV		V		VI	
		Au 197.2		*Hg* 200.6		*Tl* 204.4 *Act D* 206.5 *Th D* 208.4 RaC_2 210.5		*Pb* 206.5 ThD_2[c] 208.4 *Ra D* 210.5 *Act B* 210.5 *Th B* 212.4 *Ra B* 214.5		*Bi* 208.4 *Ra E* 210.5 *Act C* 210.5 ThC_1 212.4 RaC_1 214.5		*Ra F* 210.5 ThC_2 212.4 *RaC′* 214.5 *Act A* 214.5 *Th A* 216.4 *Ra A* 218.5
Act Em 218.5 *Th Em* 220.4 *Ra Em* 222.5	$Act X_2$[b] 218.5 $Th X_2$[b] 220.4 *Ra X*[b] 222.5											
			Act X 222.5 *Th X* 224.4 *Ra* 226.5 *Mes Th I* 228.4		*Act* 226.5 *Mes Th II* 228.4		*Rad Act* 226.5 *Rad Th* 228.4 *Io* 230.5 *Th* 232.4 *Ur X* 234.5		$Ur X_2$[a] 234.5		*Ur II* 234.5 *Ur I* 238.5	

[a] This is the hypothetical product between *Ur X* and ionium.
[b] These are the hypothetical products between *Ra*, *Th X*, and *Act X* on one side and the three emanations on the other.
[c] This is the unknown transformation product of $Th C_2$ which corresponds to *Ra D*.

O	I	II	III	IV	V	VI
— 221.5	197.2 221.5	200.3 226.5	207.5 228.5	210.5 231.5	211.5 234.5	214.5 236.5
— *Ra Em* 222.5	*Au* 197.2 (*Ra X*) 222.5	*Hg* 200.3 *Ra* 226.5	*Tl* 204.0 *Mes II* 228.4	*Pb* 207.1 *Th* 232.4	*Bi* 208.0 $Ur X_2$ 234.5	*Pol* 210.5 *Ur I* 238.5

agrees also with the tables of Mendeleyev and of Lothar Meyer in that *here also in each group one must assume two subgroups;* polonium of the upper horizontal row corresponds to tellurium of the second subgroup and not to tungsten of the first, while in the same way *RaD* is analogous to lead and not to cerium. One must assume the same in the upper horizontal row for the third and the fifth groups also. These results speak convincingly in favor of the idea that *the transformations of the elements form the basis of the periodic system.*

Now let the second arrangement be compared with the two lowest rows of the present table of the periodic system.[32] In the lower row, in both cases, one finds in the sixth, fourth and second group, the acknowledged elements *Ur*, *Th*, and *Ra*, for which direct atomic weight determinations have been carried out. The other elements which still belong in this group (see the table) are all much shorter-lived than the corresponding one of these three elements. The experimentally determined atomic weights are therefore to be considered valid when applied to the simple elements. The gaps in the fifth, third, and first groups of the lower row of the ordinary table are filled in our arrangement with short-lived, and in part hypothetical elements. There is no wonder that they were not discovered by ordinary chemical methods. The same thing occurs with polonium which fills the present gap in the sixth group of the upper horizontal row. To be noted are the consequences which appear for the elements of the third, fourth, and fifth groups of the upper row. These places are occupied in the ordinary table by the elements thallium, lead, and bismuth. In addition to these, however, there belong here several radioactive elements, as the table shows. Lead belongs to the radium series, and if we find for the atomic weight of the fourth group of the upper horizontal row not the mean value, 210.5 but that of lead, 207.1, it results from this: every other product of the group—with the exception of ThD_2, whose life we do not know—is so extraordinarily short-lived in comparison with lead that its small quantity can exert no influence on the atomic weight. The conjecture now is forced upon us that perhaps the same thing occurs for bismuth and thallium, that is that they also may be long-lived members of a transformation series. It can easily be shown that this is very probable. If the thorium series is continued by analogy with the radium series, there results:

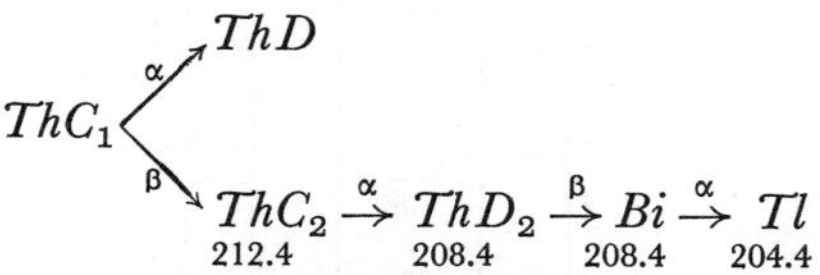

$$ThC_1 \begin{cases} \xrightarrow{\alpha} ThD \\ \xrightarrow{\beta} \underset{212.4}{ThC_2} \xrightarrow{\alpha} \underset{208.4}{ThD_2} \xrightarrow{\beta} \underset{208.4}{Bi} \xrightarrow{\alpha} \underset{204.4}{Tl} \end{cases}$$

[32] Cf. Nernst, Theoret. Chemie, 7. Aufl., 1913, p. 181. [Reprinted on p. 56 of this volume —A.R.]

The still-unknown ThD_2 would be formed, in correspondence with *RaD*, from the ThC_2 of the sixth group and would belong in the fourth group. This product must be very stable since otherwise it would have been detected by its radioactive properties, which is not the case. If one assumes that like *RaD* it supports a beta-ray transformation, by our rule an element of the fifth group (answering chemically to *RaE*) must be formed. The atomic weight of this element in fact agrees very closely with that of bismuth[33] (208.0). It is very remarkable that the atomic weight of thallium (204.0) is exactly four units smaller than that of bismuth. This speaks for an alpha-ray transformation of bismuth.[34] In the best agreement with this is the fact that in this hypothetical transformation the crossing takes place into the second group (from the fifth into the third). The presence of both thallium and bismuth in radioactive minerals as well as that of thallium in bismuth[35] could be mentioned as provisional evidence for this view. Exact analyses are needed, however, to support it better.

A discrepancy still remains to be explained. It is known that the experimentally determined atomic weight of lead (207.1) does not agree with that calculated (206.5) on the basis of the assumption that lead represents the end product of the uranium series. The same applies to the atomic weights of bismuth and thallium, only here the experimentally determined values (208.0 and 204.0) are smaller than the calculated (208.5 and 204.5). Whether or not these atomic weight

---

[33] Simply by considering atomic weights, Marsden and Darwin have already concluded that bismuth is the end product of the thorium series. Proc. Roy. Soc. **87**, A, 17, 1912.

[34] According to this conception, the *Bi* corresponding to *RaE* would be subject to an alpha-ray transformation, while *RaE* undergoes a beta-ray transformation to *RaF*. However if one assumes that a branching appears at *RaD* in this fashion:

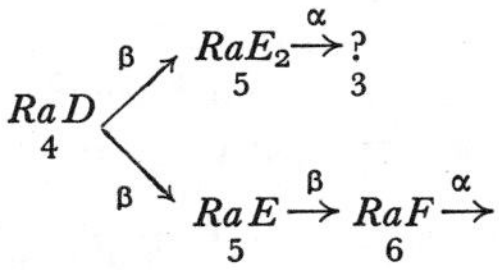

$$RaD_{4} \begin{cases} \xrightarrow{\beta} RaE_2\,(5) \xrightarrow{\alpha} ?\,(3) \\ \xrightarrow{\beta} RaE\,(5) \xrightarrow{\beta} RaF\,(6) \xrightarrow{\alpha} \end{cases}$$

then the analogy would be restored. Such an assumption could explain the not yet completely understood fact that the number of alpha particles of *RaF* is approximately twice as great as the number of beta particles of *RaE*. (Cf. Moseley, Proc. Roy. Soc. **87**, A, 230). It need only be assumed that the life of RaE_2 is close to that of *RaF* and that the branching ratio is 1:1. This assumption would also explain why polonium is similar to tellurium as well as bismuth, since it would be a mixture of RaE_2 (5th group) and *RaF* (6th group). Experiments to test this assumption are in progress.

[35] Cf. Abegg's Handbuch d. anorg. Chemie III, 1, p. 409.

determinations are to be thought inexact, the following interpretation seems not unlikely. From the preceding ideas it follows immediately that ThD_2, which has a higher atomic weight (208.4) than the lead constituted from uranium (206.5), must be very thoroughly like it. If ordinary lead represents a mixture of these two not easily separable kinds of lead, the discrepancy referred to would be explained. For *Bi* and *Tl* one must assume admixtures of products with higher atomic weight (perhaps further transformation products of lead).[36]

One question must still be discussed here, the genetic connection of the actinium series with the uranium series. It is extraordinarily likely that the actinium series represents a branching of the uranium-radium-polonium series.[37] It is still undecided, however, at what place in the series the branching appears. Actinium belongs, as has already been mentioned, to the second or third group of the periodic system. If it is in the third group, it could be constituted from an element of the fifth group by an alpha-ray transformation or from an element of the second group by a beta-ray transformation. For the first, only the hypothetical UrX_2 would come into consideration. However if this were the parent substance of actinium, one would be able to obtain actinium from uranium or *UrX* after a short time on account of its relatively short life.[38] This however, as adequate investigations by Soddy showed, is not the case. Only radium comes into consideration as a divalent element, and the beta rays detected for it could belong in part to a transformation of that kind. We should then have the transformations

$$\underset{4}{Io} \xrightarrow{\alpha} \underset{2}{Ra} \xrightarrow{\beta} \underset{3}{Act} \xrightarrow{\beta} \underset{4}{Radioact}$$

which correspond entirely to the transformations

$$\underset{4}{Th} \xrightarrow{\alpha} \underset{2}{Mes\,I} \xrightarrow{\beta} \underset{3}{Mes\,II} \xrightarrow{\beta} \underset{4}{Radioth}$$

However if actinium is a member of the second group, one must first assume the existence of a product between it and the radioactinium of the fourth group, to remain in agreement with our rules:

$$\underset{2}{Act} \xrightarrow{\beta} \underset{3}{?} \xrightarrow{\beta} \underset{4}{Radioact}$$

An element of the second group can be constituted either from one of

---

[36] To test this assumption, atomic weight determinations should be carried out on lead and bismuth which have been extracted from thorium-free uranium minerals and from uranium-free thorium minerals.

[37] See the author's Inaugural Lecture.

[38] Cf. Mme. Curie, Le Radium **8**, 353, 1911.

the fourth group by an alpha-ray transformation or from one of the first by a beta-ray transformation. For the first, ionium would come into consideration and the transformations

$$\underset{4}{Io} \xrightarrow{\alpha} \underset{2}{Act} \xrightarrow{\beta} \underset{3}{?} \xrightarrow{\beta} \underset{4}{Radioact}$$

would once more correspond entirely with those of the thorium series. According to the still possible assumption that the hypothetical alkali metal *RaX* is the parent of actinium, the analogy with the thorium series would not be so complete; the assumption is therefore less likely. If then no new kind of phenomenon appears in the formation of actinium one of the two schemes given must be assumed. The second has the advantage, as is easily perceived, that it admits an explanation for the complex radiation of radioactinium, since its beta rays would belong to the hypothetical intermediate product. It would not be difficult to decide this question experimentally. In both cases the atomic weight of actinium would be equal to that of radium, and the corresponding values of the atomic weights of the products of the actinium series are given on this supposition in the table. They possess of course only a hypothetical value.

The idea that the periodic system of the elements constitutes an expression of the evolution of matter is already old. There are many attempts to use the radioactive processes for the interpretation of the periodic law. The advance that the present paper brings in this direction consists in its having shown for the first time how in fact, from consideration of the chemical nature of the radioactive elements and their genetic connections, an arrangement is obtained which corresponds in every point with the last two rows of the periodic system. Thus it has become very probable that the periodic law is an expression of the periodic character of the radioactive transformations.

It is clear that such a result casts much new light on many questions about the periodic system. It is possible indeed on the basis of the new ideas to explain simply and without constraint many of the previous difficulties in the periodic system—for example the behavior of iodine and tellurium and the placing of the rare earth elements—and to refer the regularities observed before to a surer foundation. These questions will be discussed in another place.

Summary

1. From a law proved in the previous paper it is concluded that an alpha-ray transformation is accompanied by a crossing from right to left in a horizontal row of the periodic system. The observation of Soddy that there is a jump into the second adjacent group is assumed as generally valid.

2. In the same way it is deduced for beta-ray transformations that they produce a crossing into the next higher group, and thus from left to right in a horizontal row.

3. On the basis of these two rules, all the radioelements can be inserted into the groups of the periodic system with no contradictions.

4. An arrangement is obtained which agrees in all its properties with that to be expected for the last two rows of the periodic system. Thus it is possible to demonstrate on a sure foundation that these two rows form an expression of the laws of the transformation of the elements.

Karlsruhe i. B., Institute for Physical Chemistry of the Technical Academy.

(Received 31 December 1912).

[To go strictly by dates, Fajans had been anticipated. On January 31, Alexander S. Russell, a former student of Soddy's, published in *The Chemical News* a paper on placing the radioactive elements in the periodic table. This was two weeks before Fajans's papers appeared, although a month after he had placed them in the hands of the editor. Aside from the date, however, there is not a great deal to be said for Russell's system. He could hardly go wrong with elements for which the chemistry was known; for others he relied on his imagination. He took from Soddy the rule which moved a radioactive product two places away from its originator after the emission of an alpha particle, and added another of his own which moved it to an adjacent place after the emission of a beta particle. In both cases, however, he allowed the motion to occur either up or down the periodic table. Thus it was possible, when he came to the emanations in the zero column, which transformed by emitting alpha particles, for him to put their successors (Ra A, Th A, Ac A) in the second column in the place already occupied by mercury.[39]

It was Soddy's preparatory work which had made both Fajans's and Russell's proposals possible, partly through the useful summaries in his book, but partly also by his hypothesis of chemical identity which made it feasible to crowd some thirty known radioactive elements into the space below uranium in the periodic table. There was some justification then for Soddy to publish his own system, which is reprinted as the next paper, in spite of the genuine priorities which the other two had established. Soddy relied more than Fajans on the accumulated chemical evidence, and since he had presented a good deal of this before, his paper is shorter and less tightly argued. It is interesting to notice the easy generosity

---

[39] A. S. Russell, "The Periodic System and the Radio-Elements," *Chem. News*, 1913, *107*: 49–52.

with which he disagrees with his two predecessors: his silent omission of Fajans's radium X, and his attribution of the two displacement rules to Russell with the tacit removal of the incorrect backward motions.

There is interest also in his closing note with its suggestion that chemically identical substances might also have identical spectra. This was a reasonable extension of the hypothesis that their chemical identity did constitute them a single element. Yet as the hint of surprise suggests, it was an extension which logic could not make entirely plausible.—A. R.]

24

Frederick Soddy

The Radio-Elements and the Periodic Law

[From *The Chemical News* 1913, *107*: 97–99 (28 February).]

In the paper in which I proved that the two radio-elements, mesothorium I. and radium, are non-separable by chemical processes, and by fractional crystallisation of the chlorides, although the atomic weight of the two elements differs by about two units, it was pointed out that some of the common elements might also be mixtures of non-separable elements of different atomic weight in constant proportions. In a recent book ("Chemistry of the Radio-elements," p. 30) I stated the rule that held good in several cases, that when the α-particle was expelled the atom passed from a family of even number in the Periodic Table to the next lower-numbered even family, the family of odd number being always missed. Further, in the changes in which the α-particle was not expelled the atom in several cases reverted to its original group, resulting in a curious alternation of properties as the series proceeds. Now, when this occurs, an element of the fourth family, for example, expelling an α-particle and becoming a member of the second family, which after further changes reverts to the fourth family, the two representatives of the fourth family so resulting are not merely similar in chemical properties. They are non-separable by any known process. This applies not merely to the disintegration products of one series but to all the products. Thus, in the fourth group, thorium, uranium X, ionium, radio-thorium, radioactinium are all chemically non-separable, though they result from three separate series, and the calculated atomic weight varies from 234 to 228.

I suggested to my demonstrator, Mr. Alexander Fleck, that he should make a systematic investigation of as many of the radio-elements as possible, the chemical nature of which remained indefinite, and the first part of his results has recently been communicated to the Chemical Society (*Proc. Chem. Soc.*, 1913, xxix., 7). As the result of this work at the present time it is possible to state or predict the chemical nature of every known member of the disintegration series, and to bring these series from end to end under a few general laws of the type described, which throw a flood of new light on the nature of the Periodic Law. The lacunae that still remain to be filled up between uranium and ionium, owing to existence in this part of products with periods of the order of millions of years, can be discussed in a much narrower way in consequence.

In a paper published by A. S. Russell recently (CHEMICAL NEWS, 1913, cvii., 49) some of these generalisations have already been dealt with. Mr. Russell put forward a corollary to my rule for the α-particle which he had previously communicated to me privately in a letter in October, 1912, and which has since been strikingly verified by some of Mr. Fleck's results. Mr. Russell's rule refers to the β-ray and rayless changes, and is that when a β-ray or rayless change occurs the atom changes in chemical nature so as to pass into the family in the Periodic Table next higher in number. That is, the passage in these cases is always from an even to an odd, or from an odd to an even-numbered family. G. von Hevesy (*Phys. Zeit.*, 1913, xiv., 49), who has also been working in Prof. Rutherford's laboratory on the valency of the disintegration products, has put forward very similar views, the difference being that the effect of the β-ray change is considered by him to be the opposite or "polar" to that of the α-ray change, the valency increasing by two after a β-ray change:

The same questions are also very clearly discussed by K. Fajans, who has been connected with the development of our knowledge in the branch series, but his paper did not come to hand until after this paper was drafted (*Phys. Zeit.*, 1913, xiv., 131 and 136). He takes the view here advocated that the Periodic Law is the expression of the periodic character of radio-active changes, and anticipates some of the other points dealt with in this paper.

There is no doubt that Mr. Russell's corollary to my α-ray rule is correct, and I have adopted it just as he has put it forward, but it is possible from Mr. Fleck's results still to learn a good deal that is quite definite as to the nature of radio-active change.

In the first place let us consider the radium series from the emanation to the ultimate product with the branch series that occurs at the C-member. These branch series are now fairly clear owing to the work of Barratt, Marsden, and Darwin in the thorium series, and Makower and Fajans in the radium series, and as I have discussed them fully from the standpoint of the theory of multiple disintegration in the "Annual Report on Radio-activity for 1912" (*Ann. Reports of Progress of Chemistry*, Chemical Society, 1912), I need not further deal with them here. Apart from the definite recognition of polonium (RaF) as the homologue of tellurium, and of radium D as non-separable from lead, there was practically nothing known of the chemistry of the other members. From von Lerch's rule and the easy deposition on metals before the phenomenon had been recently investigated from the electrochemical standpoint by v. Hevesy (*Phil. Mag.*, 1912, [6], xxiii., 628) the impression had become general that they might be allied to the noble metals.

$$
\begin{array}{l}
\text{Em} \xrightarrow{\alpha} \text{A} \xrightarrow{\alpha} \text{B} \xrightarrow{\beta} \text{C} \text{ (0, VI., IV., V.)} \\
\text{C} \xrightarrow{\beta} \text{C}' \xrightarrow{\alpha} \text{D} \xrightarrow{\beta} \text{E} \xrightarrow{\beta} \text{F} \xrightarrow{\alpha} \text{G} \text{ (VI., IV., V., VI., V.)} \\
\text{C} \xrightarrow{\alpha} \text{C}_2 \xrightarrow{\beta} \text{?} \text{ (III., IV.)}
\end{array}
$$

From Fleck's results the series runs:—A, unknown; B, lead; C, bismuth; C′, unknown; D, lead; E, bismuth; F, polonium; G, lead. The known members are in accord with the rule about the expulsion of α- and β-particles, and if we extend the rule to the members the chemistry of which is still unknown, it gives for the members of the families of the successive members:—A, VI.; B, IV.; C, V.; C′, VI.; D, IV.; E, V.; F, VI.; G, IV. Now, applying the rule that when similar groups recur the elements are not merely similar, but non-separable, we can predict that in chemical behavior the two unplaced substances RaA and RaC′ will be non-separable from polonium. The prediction with regard to C′ is unverifiable, as the period of this body is estimated to be only 10^{-6} sec. But with regard to RaA it should be possible to test it, and this is now being done. As regards the branch series, which is difficult to investigate, as only some three out of 10,000 of the atoms choose this route, it may be predicted that RaC_2 is in the third family, and will prove to be non-separable from thallium, as will later be discussed. The end-product of the branch is again lead. The atomic weight of the two end-products, both non-separable from lead, are about 206 and 210 respectively.

In discussing the thorium series similarly, it must be noted that the product termed thorium D is not the analogue of radium D, but of radium C_2, as thorium D is formed from thorium C′, the product giving the longest α-rays, and therefore possessing the shortest life known (estimated as 10^{-11} second).

$$
\begin{array}{l}
\text{Em} \xrightarrow{\alpha} \text{A} \xrightarrow{\alpha} \text{B} \xrightarrow{\beta} \text{C} \text{ (0, VI., IV., V.)} \\
\text{C} \xrightarrow{\beta} \text{C}' \xrightarrow{\alpha} \text{?} \text{ (VI., IV.)} \\
\text{C} \xrightarrow{\alpha} \text{D} \rightarrow \text{?} \text{ (III., IV.)}
\end{array}
$$

THORIUM SERIES.

It is hardly necessary to discuss it in detail. Thorium A and C′ should be non-separable from polonium, though their periods are too short for this to be determinable. Both end-products, formed in this case in the proportion 65 to 35, should be non-separable from lead, the calculated atomic weight of this "lead" being about 208·5 in each case. ThD, like RaC_2, should prove to be non-separable from thallium.

The actinium series from the emanation is precisely analogous, except that no branch series has yet been established. Probably one will be found to exist, but as in radium it may represent an insignificant proportion only of the atoms disintegrating. If actinium is assumed to be analogous to thorium, again both products are non-separable from lead. The main series is analogous to that of thorium and the opposite of that of radium, actinium D being formed in the α-ray change of actinium C.

The series before the emanations furnish also, as Mr. Russell has shown, perfect illustration of the applicability of the rules with regard to α and β-changes.

$$\rightarrow \underset{\text{IV.}}{\text{Io}} \xrightarrow{\alpha} \underset{\text{II.}}{\text{Ra}} \xrightarrow{\alpha} \underset{\text{0.}}{\text{Em}} \xrightarrow{\alpha} \text{\&c.}$$

$$\underset{\text{IV.}}{\text{Th}} \xrightarrow{\alpha} \underset{\text{II.}}{\text{Msth I.}} \xrightarrow{\text{Rayless}} \underset{\text{III.}}{\text{Msth II.}} \xrightarrow{\beta} \underset{\text{IV.}}{\text{RaTh}} \xrightarrow{\alpha} \underset{\text{II.}}{\text{ThX}} \xrightarrow{\alpha} \underset{\text{0.}}{\text{Em}} \xrightarrow{\alpha} \text{\&c.}$$

$$\underset{\text{III.}}{\text{Ac}} \xrightarrow{\text{Rayless}} \underset{\text{IV.}}{\text{RaAc}} \xrightarrow{\alpha} \underset{\text{II.}}{\text{AcX}} \xrightarrow{\alpha} \underset{\text{0.}}{\text{Em}} \xrightarrow{\alpha} \text{\&c.}$$

In the thorium series Fleck has shown that mesothorium II., the only member the chemistry of which remained unknown, is non-separable from actinium. Mr. Cranston in this laboratory is just completing a research, in which he has proved that radio-thorium is the direct product of mesothorium II., a lacuna that had previously not been filled up, and which is important because the parent of ionium, the product in the uranium series corresponding with radio-thorium in the thorium series, is still experimentally unknown. The actinium series is almost absolutely analogous with the thorium series, actinium itself corresponding with mesothorium II., from which it is non-separable, the only difference being that actinium is rayless like mesothorium I., instead of giving β- and γ-rays like mesothorium II. Fleck has shown that radio-actinium, the chemistry of which had remained indefinite, is non-separable from thorium. Russell and Chadwick have described the separation of a body giving α- and β- and γ-rays from radio-actinium, but as radio-actinium is not known to give

γ-rays and only gives very soft β-rays, it seems advisable to await further results.

The radium series from ionium onwards is precisely similar to that of thorium from radio-thorium onwards, but lacunae remain in the connection between ionium and uranium. This part of the table I have been engaged in examining for the last ten years, and my results, though negative for the most part, at least exclude many possibilities. It is clear, as Mr. Russell and several others have pointed out, that uranium X must be the product of uranium I. and not of uranium II., as hitherto generally assumed. Because the product of uranium I., an α-particle being expelled, must belong to the fourth group, and for this to get back to the sixth group at uranium II., two β or rayless changes must ensue. Similarly, the unknown product of uranium X must belong to the fifth group, nowhere else represented, and be the homologue of tantalum. Calling this ekatantalum [eka Ta]* it is possible to predict a good deal about its properties from general principles, and the results of my experiments on the product of uranium X (*Phil. Mag.*, 1909, [6], xviii., 858; 1910, xx., 342). It can only give α-rays if its period is extraordinarily long. If it does not give α-rays its product must be in the sixth group. So that with fair probability we may write, as Russell does:—

$$\underset{\text{VI.}}{\text{U I.}} \xrightarrow{\alpha} \underset{\text{IV.}}{\text{UX}} \xrightarrow{\beta} \underset{\text{V.}}{\text{[eka Ta]}} \xrightarrow{\substack{\text{Rayless} \\ \text{or } \beta}} \underset{\text{VI.}}{\text{U II.}} \xrightarrow{\alpha} \underset{\text{IV.}}{\text{Io}} \xrightarrow{\alpha}, \text{\&c.}$$

Now, the α-ray producing members all belong to the even families, except in the case of the members where dual disintegration occurs. The C-members, like [eka Ta], belong to the fifth family, and undergo a dual change, such that in one branch a β-ray is followed by an α-ray change, and in the other an α-ray change is followed by a β-ray change. It is scarcely possible to resist the temptation of supposing that such a change is actually going on in [eka Ta], the rayless change into U II., followed by the α-ray change of that substance, in the one branch, proceeding simultaneously with an α-ray change into actinium, which, as is known, changes raylessly, in the other branch.

With regard to the experiments on the product of uranium X in which no growth of α-rays could be detected during the decay of the β-rays, it may be pointed out that a very small growth would be masked by the simultaneous decay of the undeviable β-rays at the same rate. Further, all the products gave α-rays after the β-radiation had disappeared. The first products possessed an activity, which consisted largely of actinium, but in the last products separated, this

* [The square brackets here and in the succeeding passages of this discussion are Soddy's.—A.R.]

constituent could only just be detected initially by the active deposit test.

$$\underset{\text{VI.}}{\text{U I.}} \xrightarrow{\alpha} \underset{\text{IV.}}{\text{UX}} \xrightarrow{\beta} \underset{\text{V.}}{[\text{eka Ta}]} \begin{cases} \xrightarrow{\text{Rayless}} \underset{\text{VI.}}{\text{U II.}} \xrightarrow{\alpha} \underset{\text{IV.}}{\text{Io}} \xrightarrow{\alpha}, \text{\&c.} \\ \xrightarrow{\alpha} \underset{\text{III.}}{\text{Ac}} \xrightarrow{\text{Rayless}} \underset{\text{IV.}}{\text{RaAc}} \xrightarrow{\alpha}, \text{\&c.} \end{cases}$$

I have some very slight evidence that actinium is really produced from uranium X through an intermediate substance. The actinium present in the last two lots of uranium X preparations, separated from 50 kilogrms. of uranium in June and September, 1909, has been accurately measured over a term of four years, and the measurements reveal an excessively minute but regular growth. In the table the activity of the active deposit produced from the preparations at different times is shown.

	Days from preparation.			
	328.	432.	829.	1281.
Activity (divisions per minute) due to actinium active deposit	1·43 (?)	1·53	1·70	1·90

The effects are excessively feeble. The conditions of the first measurement were not absolutely comparable with the others, and this is the meaning of the (?). It is clear that much further time must elapse before such measurements can be depended upon for proof of the growth of actinium. If the above scheme is correct it is certain that the period of [eka Ta] must be very long, but as no information exists as to the period of actinium it is impossible to give an estimate. In the meantime the very definite view it is now possible to take of the nature of [eka Ta] should facilitate the search for this missing element, and lead to the problem being more directly elucidated.

Without a single exception, though some cases still remain to be studied, the rule is borne out that all members in all three disintegrations series placed in the same family are non-separable from one another. When any place in the Periodic Table was before unoccupied the radio-element first discovered to be occupying it appeared as a new type. Thus the homologues of barium and tellurium respectively were unknown before radio-active investigations. Radium and polonium appeared as new types separable from barium and tellurium respectively. But the products radio-thorium, uranium X, ionium, and radio-actinium in the fourth group occupy a place already occupied by the known element thorium, and hence are non-separable from thorium.

This enables it to be predicted that eka-tantalum will be separable from tantalum. Actinium also will prove to be separable from lanthanum, as Geisel has stated, but not yet clearly demonstrated.

When the zero group is crossed the place in the sixth family, before occupied by uranium, is now occupied by polonium, in the fifth by bismuth, in the fourth by lead, in the third by thallium. That is, after passing the zero group an entirely fresh set of types is met with, which is shown in the ordinary periodic arrangement by dividing the families into A and B sub-groups. We may predict that the A members and the C′ members of the active deposits will be non-separable from polonium, and that RaC_2 and ThD, possibly AcD, will be not merely analogous to thallium but non-separable from it.

These results prove that almost every vacant place in the Periodic Table between thallium and uranium is crowded with non-separable elements of atomic weight varying over several units, and leads inevitably to the presumption that the same may be true in other parts of the table. As previously pointed out, nothing further is necessary to explain the failure of all attempts to obtain numerical relations between the atomic weights. The view that the atomic mass is a real constant fixing all the chemical and physical properties of the elements is combated most definitely by the fact that after the α-particle of mass 4 is expelled the members revert later to the original chemical type.

Finally, it may be predicted that *all* the end-products, probably six in number of the three series, with calculated atomic weights varying from 210 to 206, should be non-separable from lead; that is, should be "lead," the element that appears in the International List with the atomic weight 207·1. I should mention that Mr. Russell a year ago told me that he believed that the discrepancy between this value and that calculated (206·0) from the atomic weight of radium by the subtraction of five α-particles was due to the end-product of radium not being lead, but an element non-separable from it. It appears from the foregoing that all the end-products are probably non-separable from lead, and that "lead" is actually such a mixture as formerly, I supposed, might exist among the inactive elements.

If we suppose that all the lead in the world is produced as the end-products in the three disintegration series in constant proportion, the found atomic weight, 207·1, indicates that about half of it may result in that of thorium and the other half in that of uranium. It is, however, hardly profitable to go further in detail until the constancy of the atomic weight of lead from a variety of radio-active minerals has been experimentally tested.

A chart accompanies this paper, showing graphically the evolution of the elements through the Periodic Table from uranium to lead.

Physical Chemistry Laboratory,
Glasgow University, February 18, 1913.

Note added February 22, 1913.—I should have mentioned that if the scheme given is correct and ionium is the direct product of the α-ray change of uranium II.—and it is hardly possible to frame any alternative agreeing with the rules of the α- and β-ray changes laid down—my estimate of the life of ionium as at least 100,000 years (Royal Institute Friday Evening Lecture, March 15th, 1912), becomes definite. It is dependent only upon the absence of long-lived substances, intermediate between these two elements. It follows, therefore, from the experiments of Exner and Haschek, and A. S. Russell and Rossi, discussed fully in my "Annual Report of Radio-activity, 1913" (Section Ionium), in which these investigators failed to observe a single new line in the arc spectrum of ionium thorium preparations, that ionium and thorium must give the same arc spectrum. For on my minimum estimate of the life, the proportion of ionium in the preparations must have been at least 10 per cent by weight. The identity of the spectra of two non-separable elements, although at first sight a somewhat startling deduction, is in agreement with the modern view that the spectrum does not reveal the inner constitution of the atom, as previously assumed, but only its external characteristics. The electrons giving rise to spectra are probably those which condition valency and chemical properties, and are not what have been termed "constitutional electrons." Such "constitutional electrons" are, of course, merely a name for the material atom itself, considered apart from its electronic satellites.

I also take the opportunity of mentioning that Mr. Fleck tells me that he can effect the quantitative separation of thorium D, from the B and C members of the active deposit group, by precipitating potassium in the solution by means of platinic chloride. Taken in conjunction with the other reactions this is in conformity with the prediction that the D members of thorium and actinium will prove to be non-separable from thallium.—F. S.

[The material of the next paper has been anticipated in part. It contains the full report on Alexander Fleck's investigations into the chemical properties of uranium X, radioactinium, mesothorium 2,

III B. THALLIUM 204

IV B. LEAD 207

V B. BISMUTH 208

VI B. POLONIUM [210]

VII B. [IODINE]

ZERO. [XENON]

I A. [CAESIUM]

II A. RADIUM 226

III A. [LANTHANUM]

IV A. THORIUM 234

V A. [TANTALUM]

VI A. URANIUM 238

205 210 215 220 225 230 235 240

Tl 204

Pb 207

Bi 208

TH-D

Ra-C_2 AcD

Ra-D

Ra-E

Ra-F

Th-B

Th-C

Th-C'

Ra-B AcB

Ra-C AcC

Ra-C' AcC'

Th-A

Ra-A AcA

Th Em

Ra, Ac Em

ThX

Ra. AcX

MsTh I

MsTh II

Ra Th

Io Ra-Ac

ACTINIUM

THORIUM

UX

EKA TA

U II

URANIUM I

END RA

END TH

END Ac

END TH

END Ra, Ac

α-RAY CHANGE

β-RAY (RAYLESS) CHANGE

ALL ELEMENTS IN THE SAME VERTICAL ROW ARE NON-SEPARABLE BY CHEMICAL METHODS

thorium B, radium B, actinium B, thorium C, radium C, actinium C, and radium E. These investigations are interesting for their results, which pinned down even more firmly Fajan's and Soddy's locations of the radioactive elements in the periodic table. They are perhaps even more interesting for the methods which Fleck (and Soddy in whose laboratory the work was done) felt adequate for determining the isotopic natures of the various substances. It must be remembered that Fleck could not be content with showing merely chemical likeness. He must demonstrate that a radioactive element, present in quantities far too small to be detected by any other means than by its radiation, was so very closely similar to a common element that the two might be considered identical. To accomplish this was thoroughly strenuous work, and the account of that work which follows is hardly to be read casually.—A. R.]

25

Alexander Fleck

The Chemical Nature of Some Radioactive Disintegration Products

[From *Journal of the Chemical Society* 1913, *103*: 381–399.]

Many of the radio-elements have been shown, after close examination, to resemble other elements, and to be non-separable from them by chemical processes; thus the impossibility of separating radio-lead (radium-*D*) from lead, mesothorium-1 and thorium-*X* from radium, ionium and radio-thorium from thorium is well known (Soddy, "Chemistry of the Radio-elements"). The chemistry of the best known disintegration products is summed up by saying that they are identical with such elements as lead in the case of radium-*D*, or as radium in the case of mesothorium-1. Other such resemblances no doubt exist, and in the present paper a systematic investigation has been made of all the radio-elements of period sufficiently long for chemical examination, in the course of which it has been shown that many of the radio-elements known to resemble other elements, either radio-active or inactive, are, in fact, non-separable from them, and several new cases of the same kind have been discovered. The work detailed in this paper increases the number of disintegration products the chemistry of which can be referred to the chemistry of ordinary non-radio-active elements.

Uranium-X.

One of the methods used by Sir W. Crookes (*Proc. Roy. Soc.*, 1900, **66**, 409) when he discovered uranium-*X* was to dissolve uranyl nitrate containing a small quantity of an iron salt in excess of dilute ammonium carbonate solution, when most of the iron remained undissolved, and this contained the β-activity due to uranium-*X*. Other methods of separating this substance are known, but until recently these methods have been quite empirical. Among such methods are: (1) Shaking uranyl nitrate, crystallised from water, with ether. The uranyl nitrate dissolves mainly in the ether, whilst the aqueous layer from the water of crystallisation contains the uranium-*X*. (2) Precipitating barium sulphate in a solution of uranyl nitrate. The uranium-*X* is adsorbed by barium sulphate and can be collected with it.

In 1909 Soddy (*Phil. Mag.*, 1909, [vi], **18**, 858) collected uranium-*X* from 50 kilos. of uranyl nitrate in a highly concentrated form. The exact nature of the impurities which contained the uranium-*X* was not determined, but it was found possible to dissolve them in concentrated ammonium carbonate, and from the solution so obtained they could be fractionally precipitated by boiling. In some cases the uranium-*X* was precipitated almost entirely in the last minute fraction, and in other cases in one of the intermediate fractions. The uranium-*X* did not seem to follow any general law, but the separation was, as a rule, sharp.

In the same year Ritzel (*Zeitsch. physikal. Chem.*, 1909, **67**, 725) found that if charcoal is shaken with a solution of uranium-*X* the charcoal adsorbs the uranium-*X*. If, however, a trace of thorium sulphate is present the uranium-*X* is not adsorbed. Later Marckwald (*Ber.*, 1910, **43**, 3421) and Keetman (*Jahrb. Radioaktiv. Elektronik.*, 1909, **6**, 265) stated that they had found that they could not separate uranium-*X* from thorium, but they gave no details of their attempts. If uranium-*X* and thorium are non-separable, the general explanation of Ritzel's experiment follows at once (Soddy, T., 1911, **99**, 72).

The question of the non-separability of uranium-*X* and thorium by chemical methods was subjected to as rigorous an examination as possible. In doing this one of two methods was used: either a mixture of thorium, another element, and uranium-*X* was made, and the second element separated as completely as possible from the thorium, after which the β-activities of the two portions were measured, and the relative amounts of uranium-*X* present in each thus determined; or a mixture of thorium and uranium-*X* was made, and the thorium separated into fractions by some systematic method, after which the β-radioactivity of the fractions was measured in the electroscope, and then their content of thorium estimated in the usual way gravimetrically, by precipitating an acid solution with oxalic acid, igniting the precipitate, and weighing as thorium oxide. In this way the activity per gram of thorium oxide was obtained. The quantity of uranium-*X* was always measured by the β-radiation from a thin layer in a standard position with an electroscope which had an aluminium base 0·1 mm. thick.

Uranium-*X* was separated from uranium with iron, dissolved in dilute acid, and added to a boiling solution of sodium thiosulphate. The sulphur which is thereby precipitated does or does not contain the uranium-*X*, according as the solution is weakly or strongly acid. Thorium behaves in exactly the same manner. If a mixture of cerium, thorium, and uranium-*X* was made it was observed that the thorium, whether separated by sodium thiosulphate, hydrogen peroxide, or potassium azoimide, always contained the uranium-*X*. In a similar manner thorium, separated from zirconium—the mixture containing uranium-*X*—by dissolving both oxalates in excess of ammonium oxalate and then strongly re-acidifying the solution, contains all the β-activity.

The residues obtained by Soddy (*Phil. Mag., loc. cit.*), from which uranium-*X* had been separated, ought on this hypothesis to be free from thorium, which was found to be the case. When uranium-*X* was mixed with them and the mixture fractionated by the ammonium carbonate method, it behaved indefinitely, but when thorium was added its separation with that element was sharp.

It is perhaps worth mentioning that a convenient method for redissolving ignited thorium oxide is to fuse it with sodium peroxide. More than 95 per cent. of thorium oxide was dissolved in one operation, and no appreciable quantity of uranium-*X* remained undissolved.

Becquerel's original method of separating uranium-*X* was by adsorption by barium sulphate. If uranium-*X* and thorium are identical it is to be expected that the adsorption will be inhibited when the latter element is present. Barium sulphate precipitated in a uranium-*X* solution which contained no thorium carried down 98 per cent. of it, and if this active barium sulphate is boiled with acid no uranium-*X* passes into solution. A solution of barium, thorium, and uranium-*X* was treated with sulphuric acid, and the barium sulphate had an activity of one-third of that of the thorium remaining in solution. The filtrate contained thorium and uranium-*X*, the latter giving a β-activity of 800·9 d.p.m.[40] per gram of thorium oxide. When the barium sulphate containing one-third of the uranium-*X*

[40] This refers to divisions per minute in arbitrary units.

was boiled with nitric acid, the barium sulphate became inactive, and the filtrate contained thorium, which gave an activity due to uranium-*X* of 778·6 d.p.m. per gram of thorium oxide. The essential difference to be noted is this: if no thorium is present uranium-*X* cannot be dissolved by boiling with acids when once it has been adsorbed by barium sulphate, but if thorium is present the barium sulphate carried down with it some thorium which can be redissolved by acids, and the concentration of the uranium-*X* in the two portions of the thorium is constant. The presence of zirconium in a barium and uranium-*X* solution has no effect whatever on the adsorption by barium sulphate. The presence, however, of cerium inhibits the complete adsorption of uranium-*X* by barium sulphate, only two-thirds of the uranium-*X* being adsorbed, but there is this essential difference from the case of thorium, that the uranium-*X* contained in the precipitated sulphate cannot be redissolved by acids from it.

Boltwood (*Amer. J. Sci.*, 1908, [iv], **25**, 269) and others have used the following method of separating uranium and thorium; a mixture of the nitrates is evaporated to dryness at 110°, and then shaken with anhydrous ether. The uranium is stated to pass into solution, whilst the thorium remains undissolved. If this were so, it would be in complete accordance with the observed facts about the completeness of the separation of uranium-*X* from uranium by the ether method already mentioned. This separation of uranium and thorium has, however, been called into question, and has been the subject of some discussion (Soddy and Pirret, *Phil. Mag.*, 1910, [vi], **20**, 345; Mlle. E. Gleditsch, *Le Radium*, 1911, **8**, 256). The distribution ratio of uranium and thorium in ether, both anhydrous and containing various quantities of water, has been the subject of a detailed examination, and it has been found that the separation is only complete when small quantities of thorium are present; thus with less than 10 per cent. of thorium nitrate present, no thorium could be detected in the ethereal solution when the mixed nitrates were shaken with one volume of water and ten volumes of ether. Under the conditions of the ether-extraction process for uranium-*X* there is a relatively large quantity of ether compared with the quantity of water, and a very great weight of uranium compared with the weight of impurities present. Under these conditions no thorium would be dissolved by the ether, and it is known that no uranium-*X* is dissolved in it. From the solution of a mixture of thorium and uranium nitrates, 50 per cent. of each, it is possible to extract both thorium and uranium-*X* in the ethereal layer; thus, three successive extractions gave extracts with β-radioactivity corresponding with 10·94, 6·81, and 2·50 d.p.m. respectively, and the concentration of the uranium-*X* extracted by ether was represented by 129·7 d.p.m. per gram of thorium oxide, whilst that of the uranium-*X* remaining undissolved was 126·9 d.p.m. per gram of thorium oxide.

A quantity of uranium-*X* containing iron was dissolved, and about 3 grams of thorium nitrate were added to the solution, which was then treated with excess of ammonium carbonate until the precipitated carbonate was redissolved. This solution was then boiled for a short time, the precipitate collected, and the boiling continued. In this way ten precipitates were formed as follows:

No. of precipitate.	Activity.	Weight of ignited thorium oxalate. Gram.	Activity per gram of ignited oxalate.
1	2·35	0·0380	61·95
2	8·98	0·1711	52·49
3	1·358	0·0225	60·36
4	4·056	0·0643	63·08
5	9·58	0·1499	63·90
6	14·02	0·2214	63·31
7	18·85	0·3026	62·30
8	30·97	0·4967	62·16
9	30·80	0·5221	59·0
10	3·977	0·0701	56·90

It is evident that no alteration in the concentration of uranium-*X* and thorium has taken place. Another series of fractional precipitations was made, in which the concentration of uranium-*X* in thorium was more than ten times as great as in the above series, but again it was proved that uranium-*X* remains uniformly distributed throughout the thorium.

Experiments were now made with fractional crystallisation. Thorium acetate containing uranium-*X* was dissolved in acetic acid, and the excess of acid slowly evaporated. The first crystals that appeared were removed, and the evaporation continued. In this way three fractions were obtained, which had activities of 169·2, 160·3, and 171·3 d.p.m. per gram of thorium oxide respectively.

Uranium-*X* is one of the disintegration products that has never been volatilised. One of the methods of dissolving thorium oxide is to volatilise it in a stream of chlorine, when the anhydrous chloride collects in the cooler portions of the furnace. Thorium oxide containing uranium-*X* was treated in this way, and the product divided itself into three parts: (1) thorium, which collected in the trap outside the furnace, (2) thorium collected in the cooler parts, and (3) thorium which remained unvolatilised. It was found that in none of these portions did the activity per gram of thorium oxide vary more than 4 per cent. from the mean.

Within the errors of experiment in none of these methods has any separation of uranium-*X* from thorium been effected, although there is no known substance outside the radio-elements which would have remained with thorium through these varied chemical processes.

Radio-actinium.

The usual method employed in the separation of radio-actinium (Hahn, *Ber.*, 1906, **39**, 1605; *Physikal. Zeitsch.*, 1906, **7**, 855) is to precipitate it with sulphur by adding an actinium solution to a boiling solution of sodium thiosulphate, but Hahn has shown that its separation is very uncertain.

Stromholm and Svedberg (*Zeitsch. anorg. Chem.*, 1909, **61**, 338; **63**, 197) have shown that radio-actinium is isomorphous with thorium, and that thorium-*X* and actinium-*X* are isomorphous with barium. This does not necessarily mean that thorium and radio-actinium are non-separable elements because radium and barium are isomorphous, yet these two elements can be separated. From analogy to the thorium and the actinium series it is to be expected that radio-actinium will be identical with radio-thorium, and hence with thorium. If this is so, Hahn's thiosulphate method should become sharp and definite in the presence of the smallest quantity of thorium.

The only β-ray substance in the actinium series is actinium-*D*, but the quantity of any of the long-lived substances, actinium, radio-actinium, or actinium-*X*, may be determined by β-ray measurements; thus, if we start with actinium-*X* free from other products the activity due to actinium-*D* rises to a maximum in two hours, and then falls with a period characteristic of actinium-*X*. In the case of a pure preparation of radio-actinium the activity will rise very much more slowly owing to the time required for the production of the intermediate actinium-*X*, and after reaching a maximum should decay to zero. If θ is the time required to reach a maximum value, then:

$$\theta = \log_e \lambda_2/\lambda_1 \big/ (\lambda_2 - \lambda_1),$$

where λ_1 is the radioactive constant of radio-actinium, and λ_2 is the radioactive constant of actinium-*X*. Taking $\lambda_1{}^{-1} = 28{\cdot}1$ days, and $\lambda_2{}^{-1} = 15$ days, θ is calculated to be 20·23 days. From the form of the curves obtained in the various cases the initial purity of the preparation can be deduced quite easily.

About 0·1–0·2 gram of an actinium preparation containing lanthanum was dissolved in acid, and 20 milligrams of thorium nitrate were added. The solution was then rendered alkaline, and the precipitated actinium and thorium hydroxides were collected. The filtrate contained actinium-*X*, which was evaporated to dryness and ignited. The precipitated actinium and thorium hydroxides were redissolved in a slightly acid solution, and added to a boiling solution of sodium thiosulphate. A precipitate of sulphur is formed, and with it the thorium is precipitated as thorium hydroxide. The filtrate from this precipitate was evaporated to dryness and ignited. In this way three preparations were obtained: (1) the evaporated and ignited filtrate from the ammonia precipitation of actinium and thorium hydroxides, (2) the dried and ignited precipitate obtained by sodium thiosulphate, and (3) the evaporated and ignited filtrate from that precipitate. The second preparation will contain the thorium, and its activity should, if it contains radio-actinium, be initially zero, and then rise to a maximum in about twenty days, and finally decay again to zero. The third preparation should contain actinium free from either radio-actinium or actinium-*X*. Its β-activity should also be, initially, zero, but it should rise continually for a number of months. The three preparations were found to behave in the manner indicated (see Fig. 1). The separation of thorium from actinium was

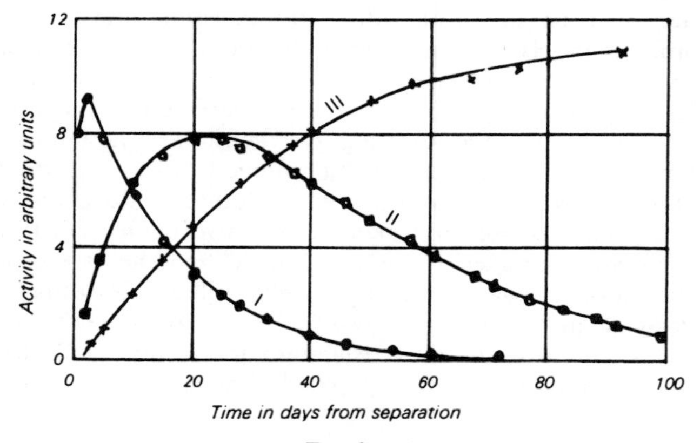

FIG. 1.

Curve I: *Activity of ammonia filtrate—Actinium-X.*
,, II: *Activity of thiosulphate precipitate—Radio-actinium.*
,, III: *Activity of thiosulphate filtrate—Actinium.*

also effected with hydrogen peroxide and potassium azoimide, and in each case the β-activity of the precipitated thorium was initially very small, but rose to a maximum in about twenty days, and then decayed to zero, with a half period of from twenty-five to thirty days,

All the available quantity of actinium was taken, and the radio-actinium separated with a small quantity of thorium by sodium thiosulphate. More thorium was added to this preparation, so that the total quantity of thorium oxide present was about 0·2 gram.

The whole was dissolved in concentrated ammonium carbonate solution, and then fractionally precipitated by boiling. The activity, measured when it had reached its maximum, per gram of thorium oxide is seen not to vary from a constant value beyond the limits of experimental error.

No. of precipitate.	Activity.	Weight of thorium oxide. Gram.	Activity per gram of thorium oxide.
1	1·512	0·0118	12·81
2	1·614	0·0136	11·86
3	2·258	0·0184	12·25
4	3·398	0·0276	12·30
5	4·565	0·0383	11·92
6	4·748	0·0412	11·53
7	4·496	0·0393	11·84
8	0·596	0·0049	12·17

It has therefore been proved that thorium and radio-actinium are alike in chemical properties so far as they have been examined, and that in place of the separation of radio-actinium from actinium being indefinite, it is effected almost perfectly if a small quantity of thorium is added to the actinium and the former separated from the latter.

Mesothorium-2

Hahn (*Physikal. Zeitsch.*, 1908, **9**, 246), who discovered this substance, recommends for its separation from mesothorium-1 the use of zirconium. This element is added to a mesothorium solution and precipitated with ammonia, when it carries down the mesothorium-2 with it. In this separation any radiothorium which is present is precipitated with the mesothorium-2.

It is easily shown that mesothorium-2 is not precipitated by hydrogen sulphide, and that therefore it must belong to the third group. It seemed at first that it might be similar to thorium, but it was observed that when thorium is precipitated with potassium azoimide or hydrogen peroxide, mesothorium-2 remains in solution. By dissolving iron, uranium, etc., with mesothorium-2 in ammonium carbonate it was found that it could be concentrated in the more insoluble fractions, and that it was unlike any of the common elements of the third group. The fact that its oxalate was very soluble suggested that it might be similar to actinium. This was tested as follows: If we start with a mixture of mesothorium-2 and actinium free from radio-actinium and subsequent products, the initial activity, if β-rays are measured, will be entirely due to mesothorium-2, which will decay almost completely to zero in two days, in which time exceedingly little actinium-*D* will have grown from actinium. In the course of three or four months actinium-*D* will reach a maximum amount, owing to the radio-actinium and actinium-*X* attaining equilibrium. Therefore, by taking decay curves during the first two days, the relative amount of mesothorium-2 can be found, and, finally, by measuring the activity after equilibrium has been reached the relative amount of actinium present can be ascertained.

Mesothorium-2 was separated from a mesothorium preparation, free from radiothorium, by means of zirconium, and added to a quantity of actinium which had been freed by sodium thiosulphate from radio-actinium (see p. 233). The whole was dissolved in ammonium carbonate and reprecipitated by boiling, and thus divided into several fractions, care being taken that the first fractions were very small. For some time the quantity of actinium-*D* in the fractions was proportional to the quantity of mesothorium-2 that had initially been present, but gradually the quantity of actinium-*D* relative to mesothorium-2 became relatively smaller in the latter fractions, but finally that quantity has again increased until now it is again almost proportional to the quantity of mesothorium originally present.

No. of precipitate.	Activity initially due to mesothorium-2.	Activity due to actinium-*D* (after $2\frac{1}{2}$ months).	$\frac{\text{Activity of actinium-}D}{\text{Activity of mesothorium-2}}$
2	6·13	8·83	1·44
3	25·01	35·31	1·41
4	7·45	9·66	1·28

Precipitates number 1, 5, 6, and 7 were inactive.

Since the quantity of actinium-*D* is finally proportional to the quantity of actinium present, it is evident that mesothorium-2 has divided itself in the same proportion as actinium. The explanation of the lack of proportionality in the earlier stages of the experiment is that the separation of actinium from thorium is not complete by one precipitation with sodium thiosulphate, and, since it has been shown that radio-actinium is identical with thorium, some radio-actinium remains with the actinium. The solubility of thorium in ammonium carbonate solution is different from that of actinium, so that radio-actinium will not be equally distributed with the actinium, and consequently there will be in the earlier stages of the experiment more actinium-*D* where radio-actinium has accumulated. It is therefore necessary if a true measure of the quantity of actinium present is required to wait until equilibrium is reached.

Another experiment was made in which actinium free from radio-actinium and actinium-*X* was dissolved in strongly acid solution along with zirconium containing mesothorium-2. Oxalic acid was added, and the solution gradually neutralised, so that the oxalates were fractionally precipitated. The result at the present time is:

No. of precipitate.	Activity initially due to mesothorium-2.	Activity due to actinium-*D* (after 2 months).	Activity of actinium-*D* / Activity of mesothorium-2
2	9·1	3·74	0·41
3	35·0	15·11	0·432
5	61·5	24·96	0·405
6	5·7	1·915	0·375
7	26·75	10·16	0·38
8	12·75	5·43	0·4255

It is to be expected that, as in the preceding experiment, the numbers in the last column will become more nearly equal as complete equilibrium is reached.

Mr. Cranston in this laboratory, who is using the method described for the separation of radiothorium from mesothorium-2, has obtained further evidence in favour of the view here taken that actinium and mesothorium-2 are chemically identical.

Chemistry of the Active Deposits.

Very little is known about the chemistry of the "active deposit group," apart from their relative volatilities in air and other gases (A. S. Russell, *Phil. Mag.*, 1912 [vi], **24**, 134), but, on the other hand, considerable progress is being made on the electrochemical side. Evidence of this is seen in von Lerch's rule, that each succeeding substance is electrochemically "nobler" than the last, and in von Hevesy's recent paper (*Phil. Mag.*, 1912, [vi], **23**, 628) on their "electrochemistry." Although so little is known there seems to be fairly general belief that they are allied to the noble metals.

Thorium-*B*, radium-*B*, and actinium-*B* give only a feebly penetrating (β)-radiation, but each gives a *C*-product which gives α-rays. If, therefore, we start with the *B*-member in a pure condition its activity will be initially zero, but will rise to a maximum, and then, since in all cases the *B*-member has a longer period than the *C*-member, the activity will decay to zero with the period of the former. If α-rays are measured, this maximum will be reached in 3·54 hours in the case of thorium-β, in 32·78 minutes in the case of radium-*B*, and in 9·29 minutes in the case of actinium-*B*, assuming the usual periods in all cases. If β-rays are measured,[41] the curves will be substantially of the

[41] This statement is not strictly true, because radium-*B* gives a small proportion of β-rays (compare Fajans and Makower, *Phil. Mag.* 1912, [vi], **23**, 293).

same form as the α-ray curves, except in the case of actinium, when the maximum for the β-rays will not be reached until a relatively greater time has elapsed, as in this case the *D*-member is longer lived than the *C*-member. If we start with the *C*-member initially pure, then the activity will decay from the beginning with a period of average life of 79, 28·1, 3·10 minutes, according as the substance is thorium-*C*, radium-*C*, or actinium-*C*.

Thorium-B *and Thorium*-C.

It was observed that when a solution of a small quantity of lead was added to a solution in which the active deposit group of the thorium series was in radioactive equilibrium, and the lead precipitated completely as sulphide, the precipitate contained thorium-*B* and -*C* in equilibrium amount. If the solution was distinctly acid, so that the lead was not completely precipitated, there was excess of thorium-*C* in the quantity of sulphide obtained. If in addition to lead the solution contained tin or other metal the sulphide of which was soluble in ammonium sulphide, then, when that sulphide was removed by this reagent, no activity was found in the solution so obtained. This shows that neither thorium-*B* nor -*C* are allied to the noble metals as hitherto supposed. If a solution of lead and copper is precipitated with sodium carbonate and the precipitate digested with potassium cyanide solution, then the copper and any other second group metal present, with the exception of lead and bismuth, will be dissolved as a complex cyanide. In these circumstances, and in the presence of thorium-*B* and-*C*, it was found that no activity was contained in the potassium cyanide solution. This shows that both thorium-*B* and -*C* are similar to lead or bismuth.

A solution of thorium-*B* and -*C* was treated with sulphuric acid and alcohol. The precipitated sulphate was removed, and hydrogen sulphide added to the filtrate to precipitate the bismuth. The activity of the lead sulphate was found to be, at the time of the first measurement, less than one-half of the value that it reached in three and a-half hours from the time of precipitation, and by producing the curves backwards from the time of the first reading until the moment of precipitation it was seen that the value then could not be more than from 8 to 12 per cent. of its maximum value. It is shown, therefore, that lead sulphate contains initially practically no thorium-*C*. On the other hand, the bismuth precipitated as sulphide gave an activity which decayed to half value in seventy-two minutes, and was therefore mainly due to thorium-*C*.

The separation of lead and bismuth containing thorium-*B* and -*C* in solution was effected in the following ways, and the activity of each precipitate measured:

(1) Lead precipitated as lead sulphate, bismuth as bismuth sulphide.
(2) Lead precipitated as lead sulphate, bismuth as bismuth hydroxide.
(3) Lead precipitated as lead chloride.
(4) Bismuth precipitated as bismuth oxychloride, lead as lead sulphide.
(5) Bismuth precipitated with *m*-nitrobenzoic acid, lead as lead hydroxide.
(6) Bismuth precipitated with alcohol.

In each of these cases it was found that the activity of the lead compound rose to a maximum, and then fell to half value in eleven hours, whilst the activity of the bismuth precipitate fell to half value with a period slightly longer than the usual value of thorium-*C*, as a rule about seventy-four to seventy-six minutes. The separation of the two metals was also effected electrolytically by depositing metallic bismuth on the cathode, and lead peroxide on the anode from an acid solution. The activity of the cathode deposit fell normally with the period of thorium-*C*, but the activity of the anode deposit rose only slightly or not at all before falling with the period of thorium-*B*. On analysis it was found that the lead peroxide contained considerable quantities of bismuth, and that, therefore, it was not possible to deposit lead and bismuth simultaneously in a pure condition.

The reaction mentioned above (No. 5) of separating bismuth by the use of *m*-nitrobenzoic acid is especially interesting. It was suggested by the work of Schlundt and Moore (*J. Physical Chem.*, 1905, **9**, 682), who found that thorium-*B* and thorium-*X* remained in solution, whilst thorium and the other products were precipitated by this reagent. Since it appears that thorium-*B* behaves as lead in all the reactions that have been tried, it was of interest to see if the converse is true, if lead will behave as thorium-*B* and will remain in solution, whilst bismuth, if it behaves as thorium-*C*, will be precipitated. This was tested and found to be the case, and a separation of these two metals could be made by this means. The precipitation of bismuth, either by *m*-nitrobenzoic acid or by alcohol, does not seem to be generally known, and it is intended to continue the examination of these two reactions to see if they are capable of application to analytical work.

Experiments were then made in the systematic fractionation of lead in a solution containing thorium-*B* to see if any difference of concentration of thorium-*B* in the various fractions could be noticed.

Lead was fractionally precipitated by adding successive small quantities of sulphuric acid and finally alcohol, the precipitate formed being removed between each addition. The activities of the various fractions were measured for two days, and then the quantity of lead sulphate in each fraction was weighed. By dividing the β-activity at any time, after the maximum is reached, measured by the ordinate at that time, by the weight of lead sulphate, the concentration of thorium-*B* in lead sulphate at that time is obtained.

No. of precipitate.	Weight of lead sulphate. Gram.	Activity at 5 p.m.	Activity per gram of lead sulphate at 5 p.m.	Activity at 11 a.m. next day.	Activity per gram at 11 a.m. next day.
1	0·1161	0·90	7·25	0·29	2·496
2	0·2879	2·086	7·26	0·68	2·360
3	0·1810	1·28	7·09	0·374	2·064
4	0·0894	0·65	7·29	0·225	2·51
5	0·0501	0·456	9·1	0·15	2·99

The last two columns are merely confirmatory evidence of the constant distribution of thorium-*B* throughout the lead. Lead was fractionally crystallised as chloride, and also fractionally volatilised as chloride, and in each case similar series to the one shown above were obtained, showing that it was not possible to alter the concentration of thorium-*B* in lead.

Thorium-*C* was fractionally precipitated with bismuth in two ways. The bismuth was dissolved in concentrated ammonium carbonate, and then reprecipitated by boiling. In every experiment that was made a very small concentration from the bismuth was observed in the opposite sense to that which would have occurred with polonium. The second method was by fractional precipitation of the oxychloride, and again a very small concentration was observed, thorium-*C* being more soluble than bismuth. The concentration of the first portion was 1·735 units of activity per gram of bismuth oxide, whilst that of the second portion was 1·88. The difference in concentration is very small, and is probably capable of the simple explanation that there is present a small quantity of thorium-*B* which produces thorium-*C* in the interval between the precipitations.

Radium-B *and Radium*-C.

In the same way as has been shown with thorium-*B* and -*C*, radium-*B* and -*C* were found to be similar to lead and bismuth respectively, to the exclusion of all other elements. A solution of the active deposit of rapid change, containing in addition lead and bismuth, was treated in the following ways so that these two metals were separated:

(1) Lead precipitated as lead sulphate, bismuth precipitated as bismuth sulphide.

(2) Lead precipitated as chloride.

(3) Bismuth precipitated as bismuth oxychloride, lead as lead hydroxide.

(4) Bismuth precipitated with *m*-nitrobenzoic acid.

(5) Bismuth precipitated as metal on the cathode, lead precipitated as lead peroxide on the anode.

The curves obtained (see Fig. 2) show that the activity of the

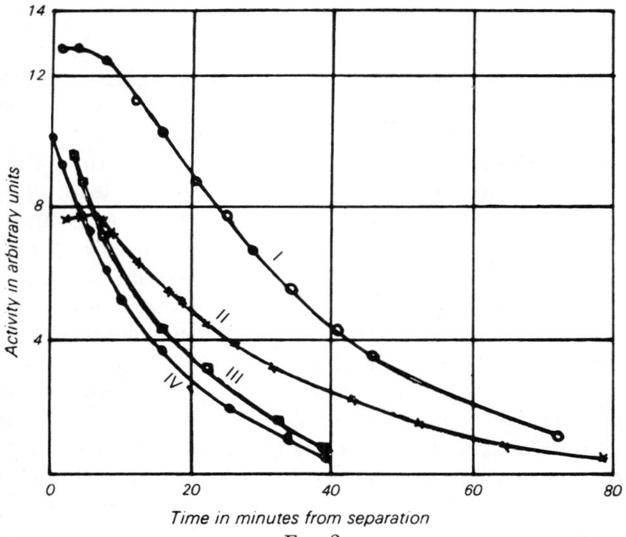

FIG. 2

Curve I: *Activity obtained in lead sulphate—Radium*-B.
" II: *Activity of lead peroxide on anode—Radium*-B.
" III: *Activity of bismuth on cathode—Radium*-C.
" IV: *Activity of ppt. with* m-*nitrobenzoic acid—Radium*-C.

lead precipitates decay to half value in thirty-six minutes, whilst the bismuth precipitates decay to half value in from twenty to twenty-two minutes, showing that the activities, in the first case, were due to radium-*B*, and in the second case, to radium-*C*. If the preparation of radium-*B* is pure initially, the activity ought to rise to a maximum in 32·87 minutes after separation. The form of the curves showed that the activities of the substances contained in the lead was falling from a maximum value, but even by precipitating lead sulphate, collecting it, and drying it within ten minutes it was not possible to get a very definite rise to a maximum.

Owing to the shortness of the periods of both substances it was found impossible to make determinations of their behaviour on fractional treatment.

The results of these experiments are therefore that in all the reactions which it was possible to try it was observed that radium-*B* behaves as lead, and that radium-*C* behaves as bismuth. Considering the known similarity between the corresponding members of the active deposit groups, it is most probable that radium-*B* and radium-*C* are chemically non-separable from lead and bismuth in the same way as thorium-*B* and thorium-*C* have been shown to be.

Actinium-B *and Actinium*-C.

In the case of these two substances the same experiments were made as with radium-*B* and radium-*C*, and it was observed that similarly they were like lead and bismuth respectively.

The activity of the lead precipitated from a solution of actinium active deposit either as lead sulphate, lead chloride, lead sulphide, or lead peroxide on the anode decayed to half value in thirty-eight minutes. In this case the activity of a lead precipitate, if it, initially, contains pure actinium-*B*, should reach a maximum value in 9·29 minutes. It was not possible to observe this maximum, and all the precipitates obtained were, by the time they could be dried and measured, decaying exponentially.

The decay of actinium-*C* could be measured in bismuth obtained as a precipitate with *m*-nitrobenzoic acid or by deposition of metallic bismuth on the cathode. Bismuth which was obtained in this way had an activity which decayed to half value in something less than three minutes. These measurements were made with α-rays. Enough actinium was not available to obtain a sufficient quantity of the products for β-ray measurements, so that it is, at the moment, not possible to say anything about the chemical behavior of actinium-*D*.

The conclusions that were drawn with respect to radium-*B* and -*C* are equally applicable here, namely, that in all the reactions that have been tried no evidence of the separation of actinium-*B* and actinium-*C* from lead and bismuth has been obtained, and it is probable that they are chemically similar to these elements.

Radium-E.

Radium-*E* was for a long time considered to be a complex substance owing to the work of Rutherford (*Phil. Mag.*, 1904 [vi], **8**, 636) and of Meyer and von Schweidler (*Sitzungsber. K. Akad. Wiss. Wien*, 1906, ii*a*, **115**, 697), but later work due to Antonoff (*Phil. Mag.*, 1910, [vi], **19**, 825) has shown it to be a single substance. This work shows that a substance with a period of average life of 7·25 days followed radio-lead, but no evidence has yet been given to show whether polonium is the direct product of radium-*E*.

A sample of radio-lead containing large quantities of lead was precipitated with sulphuric acid. By one precipitation radio-lead is freed from about 80 per cent. of its β-ray activity. By dissolving this sulphate in ammonium acetate, reprecipitating with hydrogen sulphide, redissolving in acid, and finally reprecipitating with sulphuric acid, radio-lead is obtained practically free from β- or α-ray activity. Measurements were then made with the apparatus used by Soddy (*Phil. Mag.*, 1909, [vi], **18**, 858) to attempt to detect a growth of α-radiation from uranium-*X*, in which the powerful β-radiation was suppressed by placing the preparation between the poles of a magnet. The electroscope was of brass, and had in its base an opening 9 cm. long by 1·75 cm. broad, which was covered with aluminium foil 0·0031 mm. thick. Across this opening, arrangements were made for pushing over or withdrawing a piece of mica 0·05 mm. thick. The preparations were on a thin layer of filter paper, contained in a tray 8 × 1·5 cm., which was placed between the poles of the electromagnet, and was 0·5 cm. below the aluminium foil. If the magnet was off and the mica foil across the opening, all the α-rays would be absorbed, and only β-rays and a small quantity of γ-rays would reach the ionising chamber. If the magnet was on and the mica away from the opening, most of the ionising effect would be due to α-rays, but there would be a small effect due to undeviated β-rays and the small quantity of γ-rays. The growth of α-and β-rays from two preparations of radio-lead was measured with (1) radio-lead precipitated twice as lead sulphate, and (2) radio-lead crystallised twice as lead chloride. The curves agree with those calculated from the view that the only intermediate member between radium-*D* and radium-*F* is radium-*E* with a period of 7·25 days.

The simultaneous decay of radium-*E* and growth of polonium was also observed. The purest radium-*E* was obtained by separating by two crystallisations of lead chloride both radium-*E* and polonium from a relatively large quantity of radio-lead. The separation was then repeated after two days had elapsed, and the last fractions of lead were removed as sulphate. A rod of bismuth was then immersed in the filtrate for three or four hours. In this way a strong β-ray preparation with an α-activity less than one-tenth of the β-ray activity was obtained. The β-ray activity decayed to half value in five days, whilst the α-ray activity increased according to the $I_t/I_\infty = 1 - e^{-\lambda t}$ law, showing that the α-ray substance was produced directly from radium-*E*. There is thus no evidence of any intermediate substances except radium-*E* between radio-lead and polonium (see Fig. 3).

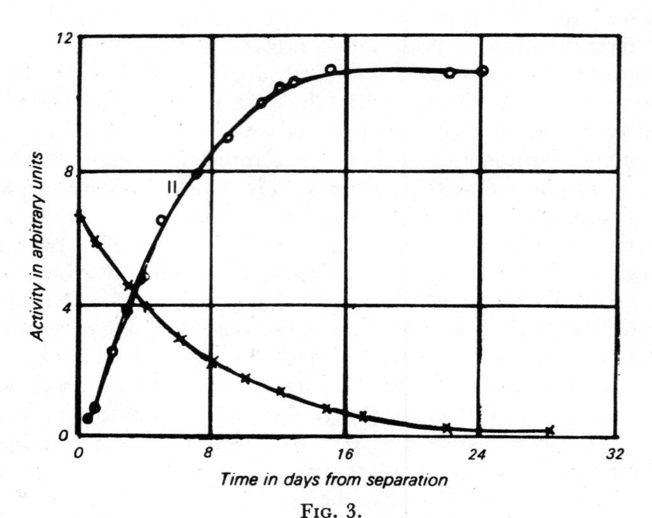

FIG. 3.

Curve I: *Decay of β-rays of radium*-E.
,, II: *Growth of α-rays of radium*-F.

(Note.—Curve II is on a larger scale than Curve I.)

The chemical nature of radium-*E* was then studied, and it was found to be in all respects similar to bismuth. It was found to remain with the bismuth when that substance was precipitated, after the removal of the lead from a radio-lead solution, by hydrogen sulphide, ammonia, or dilute ammonium carbonate, and with the bismuth when that element was precipitated from a radio-lead solution by *m*-nitrobenzoic acid or as bismuth oxychloride. Fractional precipitation experiments were also made by precipitating, by boiling, bismuth carbonate dissolved in concentrated ammonium carbonate. The decay curves were taken, and then the bismuth estimated gravimetrically as oxide.

No. of precipitate.	Weight of bismuth oxide. Gram.	Activity of precipitate on October 23rd.	Activity per gram on that date.	Activity of precipitate on October 29th	Activity per gram on that date.
1	0·1399	3·732	2·67	1·536	1·099
2	0·0920	2·377	2·58	1·033	1·120
3	0·0496	1·323	2·72	0·6054	1·215
4	0·0125	0·3356	2·68	0·1356	1·084

The last two columns confirm the regular decay of radium-*E*.

The conclusions of Rutherford already referred to concerning the complex nature of radium-*E* were arrived at by a consideration of the activities shown by the volatilised and unvolatilised parts of radium-*E* collected from radium emanation on a platinum wire. Experiments were therefore made in the volatilisation of bismuth compounds, for example, bismuth sulphide, bismuth chloride, and bismuth iodide. A strong preparation of radium-*E* had a little bismuth added to it, the whole being dissolved, precipitated as carbonate, and ignited in the filter paper, so that the oxide was reduced to metal. This was thoroughly ground with a quantity of iodine, placed at the bottom of a long, hard glass test-tube, and heated. Bismuth iodide was formed, which volatilised and condensed over a considerable length of tube. The tube was then divided into a number of sections. One section had a particularly strong activity, and the volatilisation process was repeated for it alone. In this way eight fractions were obtained, the radiation of which without exception decayed to half value in

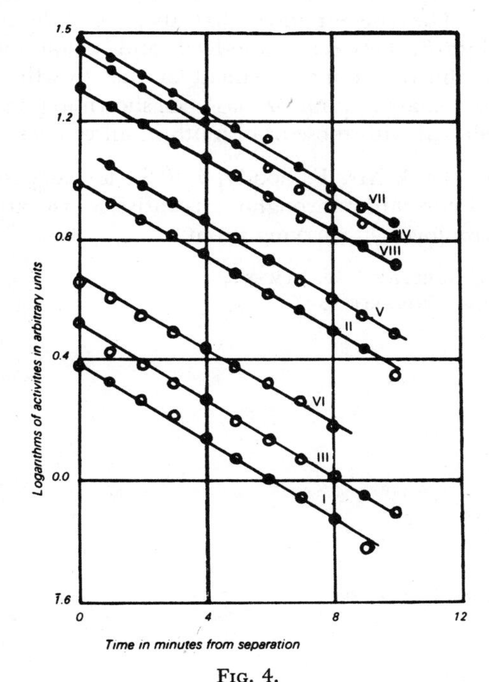

FIG. 4.

*Lines are logarithmic decay curves of radium-*E*, and are numbered in the order of the volatilities.*

from 5·0 to 5·2 days. The logarithmic curves are given in Fig. 4. As the total quantity of bismuth was equivalent to only about 0·05 gram of bismuth oxide, the amount of bismuth in each fraction was very small, and thus it was not possible to get an exact determination of the concentration of radium-*E*, but there was no distinct alteration in the proportion of radium-*E* to bismuth in any of the fractions. Had there been any other substance present, such as a second intermediate substance between radio-lead and polonium, it would have shown itself in a difference between the decay curves of the first and last fractions. No such difference was observed.

Summary.

It has been shown that uranium-*X* and radio-actinium are chemically similar to, and non-separable from, thorium; mesothorium-2 is non-separable from actinium; thorium-*B* is non-separable from lead; radium-*B* and actinium-*B* are extremely similar to lead, and most probably non-separable from it; thorium-*C*, radium-*C*, and actinium-*C* are very closely allied to bismuth, and probably chemically similar to it. The present view that there is only one product, namely, radium-*E*, between radio-lead and polonium, has been confirmed by the direct measurement of the growth of radium-*F* from radium-*E*; finally, radium-*E* has been shown to possess chemical properties identical with those of bismuth in all respects.

I desire to thank Mr. F. Soddy, F.R.S., for suggesting this research, for his interest in it throughout its entire course, and for the use of materials employed in carrying it out.

Physical Chemistry Department,
Glasgow University.

VIII. Epilogue

[It is interesting to observe how far a science can progress while its fundamental questions remain unanswered. For a decade and a half, the essential nature of the radioactive substances had been quite unknown. Whether they might be considered elements as the chemists defined an element was simply a matter of opinion. For only three, uranium, thorium, and radium, had an atomic weight been determined or a characteristic spectrum established. The transience of their existence was enough to cast doubt on all the rest, quite aside from their lack of well-defined chemical properties. Now Fajans and Soddy had managed to fit the whole group into the periodic table, and Fleck had shown that an adequate number of them possessed exactly the chemical properties appropriate to the places they occupied.

They were elements if one could recognize the internal consistency of the system or master the details of Fleck's reactions, although there were some who found such evidence difficult to accept. It was hard to believe that thorium and ionium (for example) should be called *identical* in chemistry or that atoms could be chemically inseparable if they differed in weight.[1] Nevertheless, the new ideas proved to be fruitful, and they became intellectually more manageable when Soddy coined a name for those atoms which were physically distinguishable although chemically alike, proposing to call them *isotopes*. The note in which that name first appears is printed as the last paper in this collection, and it is interesting to see how casually a word was introduced which has become a part of the standard vocabulary of physics and chemistry.

Soddy's paper brings this volume to an end in 1913, but it may be worth sketching briefly some of the investigations of the next few years which the theory of isotopes suggested.

Fajans's and Soddy's arrangements of the radioactive elements in the periodic table provided a number of predictions of chemical properties which could be tested in continuations of Fleck's work.[2] Those schemes predicted also that the stable end-products of the uranium and thorium

---

[1] A. Schuster, "The Radio-Elements and the Periodic Law," *Nature, Lond.*, 1913, *91*: 30–31.

[2] W. Metzener, "Zur Kenntnis der chemischen Eigenschaften von Thorium C und Thorium D," *Ber. dtsch chem. Ges.*, 1913, *46*: 979–986.

K. Fajans and P. Beer, "Über die chemische Natur einiger kurzlebiger Radioelemente," *Naturwissenschaften*, 1913, *1*: 338–339.

K. Fajans and O. Göhring, "Über die komplexe Natur des Ur X," *Naturwissenschaften*, 1913, *1*: 339.

A. Fleck, "The Chemical Nature of some Radioactive Disintegration Products. Part II," *J. Chem. Soc.*, 1913, *103*: 1052–1061.

series should both be isotopes of lead, the uranium-lead with an atomic weight of 206, the thorium-lead with an atomic weight of 208. Over geological time, enough of these isotopes had accumulated in radioactive minerals to be handled by ordinary chemical methods, and between 1914 and 1918 a series of atomic weight determinations on lead from very pure minerals of uranium and thorium verified the predictions of the theory.[3] The most notable of these were the elaborate investigations by T. W. Richards of Harvard on uranium-lead,[4] and the painstaking work on thorium-lead by O. Hönigschmid in Prague and in Munich.[5] Hönigschmid used specimens furnished him by Soddy (now at Aberdeen) through the intermediary of R. W. Lawson, an English physicist interned in Vienna by the outbreak of war in 1914 and working on problems of radioactivity at the university there.[6]

By the end of 1912, J. J. Thomson at Cambridge had developed an electromagnetic method for separating positive ions of different masses in a gas-discharge tube. Early in 1913 he announced that when neon was present he obtained atoms of mass 20 (corresponding to the known atomic weight of neon) and also of mass 22.[7] F. W. Aston of Cambridge

---

[3] M. Curie, "Sur les écarts de poids atomiques obtenus avec le plomb provenant de divers minéraux," *C. R. Acad. Sci., Paris*, 1914, *158*: 1676–1679.

O. Hönigschmid and S. Horowitz, "Sur le poids atomique du plomb de la pechblende," *C. R. Acad. Sci., Paris*, 1914, *158*: 1796–1798.

F. Soddy and H. Hyman, "The Atomic Weight of Lead from Ceylon Thorite," *J. chem. Soc.*, 1914, *105*: 1402–1408.

F. Soddy, "The Density of Lead from Ceylon Thorite," *Nature, Lond.*, 1915, *94*: 615.

Thomas R. Merton, "The Spectra of Ordinary Lead and Lead of Radioactive Origin," *Proc. Roy. Soc., Lond. A*, 1915, *91*: 198–201.

[4] T. W. Richards and M. E. Lembert, "The Atomic Weight of Lead of Radioactive Origin," *J. Amer. chem. Soc.*, 1914, *36*: 1329–1344; "Poids atomique du plomb d'origine radioactive," *C. R. Acad. Sci., Paris*, 1914, *159*: 248–250.

T. W. Richards and C. Wadsworth, "The Density of Lead from radioactive Minerals," *J. Amer. chem. Soc.*, 1916, *38*: 221–227; "Density of Radio-Lead from Pure Norwegian Cleveite," *ibid.*, pp. 1658–1660; "Further study of the Atomic Weight of Lead of Radioactive Origin," *ibid.*, pp. 2613–2622.

T. W. Richards and N. F. Hall, "Attempt to Separate the Isotopic Forms of Lead by Fractional Crystallization of the Nitrate," *J. Amer. chem. Soc.*, 1917, *39*: 531–541.

T. W. Richards and W. C. Schumb, "The Refractive Index and Solubilities of the Nitrates of Lead Isotopes," *J. Amer. chem. Soc.*, 1918, *40*: 1403–1409.

T. W. Richards, "The Problem of Radioactive Lead" (Presidential Address, American Association for the Advancement of Science), *Science*, 1919, *49*: 1–11; *Nature, Lond.*, 1919, *103*: 74–78, 93–96.

[5] O. Hönigschmid, "Über das Thoriumblei," *Z. Elektrochem.* 1917, *23*: 161–165; "Neuere Atomgewichtsbestimmungen (Thoriumblei, Skandium)," *ibid.*, 1919, *25*: 91–97.

[6] F. Soddy, "The Atomic Weight of 'Thorium' Lead," *Nature, Lond.*, 1916–17, *98*: 469.

[7] Sir J. J. Thomson, "Some Further Applications of the Method of Positive Rays," *Nature, Lond.*, 1913, *91*: 333–337; "Positive Rays of Electricity," *ibid.*, p. 362.

suspected that these might represent isotopes of neon and succeeded before the summer was out in separating a neon specimen by a diffusion process into two fractions of slightly but perceptibly different density.[8] Further work in this direction was interrupted by the war, and it was not until 1919 that Aston was able to begin the development of his mass-spectrograph from Thomson's discharge tubes. With this instrument he was able to show the existence of a multitude of isotopes of the common elements.[9]

In addition to its introduction of the term *isotope*, the paper of Soddy's which appears next is interesting for his adoption of the hypothesis of atomic number. This is the hypothesis that the nuclear charges of successive elements in the periodic table differ by only single units of electronic charge, and that every integral number of nuclear charge is represented there. With the mention of this hypothesis, we must introduce Antonius Johannes van den Broek (whose name Soddy misspells here), an amateur physicist, although trained as a lawyer, whose hobby was the periodic table. In 1907, he had picked up a casual remark of Rutherford's and decided that all elements must be built up of half-alpha-particles with an atomic weight of 2. To validate his idea, he had rearranged the periodic table into fifteen rows of eight elements each, which brought uranium in as the hundred and twentieth element with an atomic weight of 240, a reasonable approximation to the current value of 238.5. What was more, when he added the experimentally determined atomic weights of the known elements, he obtained a sum of 7723.65, while his system—which took the atomic weight to be twice the place number—gave for those same elements a sum of 7728, an agreement to 0.06 per cent.[10] These ideas he developed over the next half-dozen years.[11]

Rutherford's nuclear model of the atom was well adapted to van den Broek's scheme. In the paper which Soddy cites here, van den Broek was concerned to show that Geiger and Marsden's scattering data were completely consistent with his assignment of atomic numbers.[12]

As a matter of fact, Bohr had already adopted van den Broek's idea in the second of his papers on the nuclear atom, which had appeared the previous September. There he had given the same arguments that Soddy details here, to show how such an assignment of nuclear charge

---

[8] F. W. Aston, "New Element in the Atmosphere," *Engineering*, 1913, *96*: 423.

[9] F. W. Aston, *Isotopes*, New York: Longmans, Green, 1923.

[10] A. van den Broek, "Das α-Teilchen und das periodische System der Elemente," *Ann. Phys., Lpz* [4], 1907, *23*: 199–203.

[11] A. van den Broek, "Das Mendelejeffsche ‚kubische' periodische System der Elemente und die Einordnung der Radioelemente in dieses System," *Phys. Z.*, 1911, *12*: 490–497; "The Number of Possible Elements and Mendeléeff's 'Cubic' Periodic System," *Nature, Lond.*, 1911, *87*: 78; "Die Radioelemente, das Periodische System und die Konstitution der Atome," *Phys. Z.*, 1913, *14*: 32–41.

[12] A. van den Broek, "Intra-Atomic Charge," *Nature, Lond.*, 1913–14, *92*: 372–373.

accounts for the displacement of an element in its radioactive transformation.[13] By November, when Soddy was writing, H. G. J. Moseley had already acquired the notion from Bohr; and at the beginning of December he published his measurements on the x-ray spectra of the elements from calcium to zinc.[14] It was evident that the frequencies of these x-ray lines depended upon a series of successive integral numbers; and if one accepted the details of Bohr's atomic theory, those numbers represented the nuclear charges of the atoms and were equal to the place numbers of the elements in the periodic table.

The chemical evidence and physical evidence were converging beautifully.—A. R.]

26

Frederick Soddy

Intra-atomic Charge

[From *Nature*, 1913–14, *92*: 399–400 (4 December).]

That the intra-atomic charge of an element is determined by its place in the periodic table rather than by its atomic weight, as concluded by A. van der Broek (NATURE, November 27, p. 372), is strongly supported by the recent generalisation as to the radio-elements and the periodic law. The successive expulsion of one α and two β particles in three radio-active changes in any order brings the intra-atomic charge of the element back to its initial value, and the element back to its original place in the table, though its atomic mass is reduced by four units. We have recently obtained something like a direct proof of van der Broek's view that the intra-atomic charge of the nucleus of an atom is not a purely positive charge, as on Rutherford's tentative theory, but is the difference between a positive and a smaller negative charge.

Fajans, in his paper on the periodic law generalisation (*Physikal. Zeitsch.*, 1913, vol. xiv., p. 131), directed attention to the fact that the changes of chemical nature consequent upon the expulsion of α and β particles are precisely of the same kind as in ordinary electrochemical changes of valency. He drew from this the conclusion that radio-

---

[13] N. Bohr, "On the Constitution of Atoms and Molecules. Part II—Systems Containing Only a Single Molecule," *Phil. Mag.* [6], 1913, *26*: 476–502.

[14] H. G. J. Moseley, "The High-Frequency Spectra of the Elements," *Phil. Mag.* [6], 1913, *26*: 1024–1034.

active changes must occur in the same region of atomic structure as ordinary chemical changes, rather than with a distinct inner region of structure, or "nucleus," as hitherto supposed. In my paper on the same generalisation, published immediately after that of Fajans (*Chem. News*, February 28), I laid stress on the absolute identity of chemical properties of different elements occupying the same place in the periodic table.

A simple deduction from this view supplied me with a means of testing the correctness of Fajans's conclusion that radio-changes and chemical changes are concerned with the same region of atomic structure. On my view his conclusion would involve nothing else than that, for example, uranium in its tetravalent uranous compounds must be chemically identical with and non-separable from thorium compounds. For uranium X, formed from uranium I by expulsion of an α particle, is chemically identical with thorium, as also is ionium formed in the same way from uranium II. Uranium X loses two β particles and passes back into uranium II, chemically identical with uranium. Uranous salts also lose two electrons and pass into the more common hexavalent uranyl compounds. If these electrons come from the same region of the atom uranous salts should be chemically non-separable from thorium salts. But they are not.

There is a strong resemblance in chemical character between uranous and thorium salts, and I asked Mr. Fleck to examine whether they could be separated by chemical methods when mixed, the uranium being kept unchanged throughout in the uranous or tetravalent condition. Mr. Fleck will publish the experiments separately, and I am indebted to him for the result that the two classes of compounds can readily be separated by fractionation methods.

This, I think, amounts to a proof that the electrons expelled as β-rays come from a nucleus not capable of supplying electrons to or withdrawing them from the ring, though this ring is capable of gaining or losing electrons from the exterior during ordinary electrochemical changes of valency.

I regard van der Broek's view, that the number representing the net positive charge of the nucleus is the number of the place which the element occupies in the periodic table when all the possible places from hydrogen to uranium are arranged in sequence, as practically proved so far as the relative value of the charge for the members of the end of the sequence, from thallium to uranium, is concerned. We are left uncertain as to the absolute value of the charge, because of the doubt regarding the exact number of rare-earth elements that exist. If we assume that all of these are known, the value for the positive charge of the nucleus of the uranium atom is about 90. Whereas if we make the more doubtful assumption that the periodic table runs regularly, as regards numbers of places, through the rare-earth group, and that between barium and radium, for example, two complete long periods exist, the number is 96. In either case it is appreciably less than 120, the number were the charge equal to one-half the atomic weight, as it would be if the nucleus were made out of α particles only. Six nuclear electrons are known to exist in the uranium atom, which expels in its changes six β rays. Were the nucleus made up of α particles there must be thirty or twenty-four respectively nuclear electrons, compared with ninety-six or 102 respectively in the ring. If, as has been suggested, hydrogen is a second component of atomic structure, there must be more than this. But there can be no doubt that there must be some, and that the central charge of the atom on Rutherford's theory cannot be a pure positive charge, but must contain electrons, as van der Broek concludes.

So far as I personally am concerned, this has resulted in a great clarification of my ideas, and it may be helpful to others, though no doubt there is little originality in it. The same algebraic sum of the positive and negative charges in the nucleus, when the arithmetical sum is different, gives what I call "isotopes" or "isotopic elements," because they occupy the same place in the periodic table. They are chemically identical, and save only as regards the relatively few physical properties which depend upon atomic mass directly, physically identical also. Unit changes of this nuclear charge, so reckoned algebraically, give the successive places in the periodic table. For any one "place," or any one nuclear charge, more than one number of electrons in the outer-ring system may exist, and in such a case the element exhibits variable valency. But such changes of number, or of valency, concern only the ring and its external environment. There is no in- and out-going of electrons between ring and nucleus.

FREDERICK SODDY.

Physical Chemistry Laboratory,
University of Glasgow.

Appendix

TRANSFORMATION SERIES OF THE NATURAL RADIOACTIVE ELEMENTS: URANIUM, THORIUM, AND ACTINIUM SERIES

[The half-lives in this table are taken from *Chart of the Nuclides*, Knolls Atomic Power Laboratory (General Electric Co., Schenectady, N.Y.), 8th Edition, 1965—A. R.]

GENETIC NAME	ATOMIC NUMBER	MODERN DESIGNATION	RADIATION EMITTED	HALF-LIFE
		Uranium-Radium Series		
Uranium I	92	U^{238}	α	4.51×10^9 yr.
Uranium X_1	90	Th^{234}	β	24.10 days
Uranium X_2	91	Pa^{234}	β	1.18 min.
Uranium II	92	U^{234}	α	2.48×10^5 yr.
Ionium	90	Th^{230}	α	7.6×10^4 yr.
Radium	88	Ra^{226}	α	1620 yr.
Radon (Emanation)	86	Rn^{222}	α	3.823 days
Radium A	84	Po^{218}	α	3.05 min.
Radium B	82	Pb^{214}	β	26.8 min.
Radium C	83	Bi^{214}	$\alpha + \beta$	19.7 min.
($\downarrow\beta$) Radium C′	84	Po^{214}	α	1.64×10^{-4} sec.
($\downarrow\alpha$) Radium C″	81	Tl^{210}	β	1.3 min.
($\downarrow\beta$, $\downarrow\alpha$) Radium D	82	Pb^{210}	β	22 yr.
Radium E	83	Bi^{210}	β	5.0 days
Radium F (Polonium)	84	Po^{210}	α	138.40 days
Radium G	82	Pb^{206}	stable	—
		Thorium Series		
Thorium	90	Th^{232}	α	1.41×10^{10} yr.
Mesothorium 1	88	Ra^{228}	β	5.7 yr.
Mesothorium 2	89	Ac^{228}	β	6.13 hr.

GENETIC NAME	ATOMIC NUMBER	MODERN DESIGNATION	RADIATION EMITTED	HALF-LIFE
Radiothorium	90	Th^{228}	α	1.91 yr.
Thorium X	88	Ra^{224}	α	3.64 days
Thoron (Emanation)	86	Rn^{220}	α	56 sec.
Thorium A	84	Po^{216}	α	0.15 sec.
Thorium B	82	Pb^{212}	β	10.64 hr.
Thorium C	83	Bi^{212}	$\alpha + \beta$	60.6 min.
↓β Thorium C′ (↓α Thorium C″)	84	Po^{212}	α	0.30×10^{-6} sec.
Thorium C″	81	Tl^{208}	β	3.1 min.
↓β (from C″), ↓α (from C′) Thorium D	82	Pb^{208}	stable	—

Actinium Series

GENETIC NAME	ATOMIC NUMBER	MODERN DESIGNATION	RADIATION EMITTED	HALF-LIFE
Actinium U	92	U^{235}	α	7.13×10^{8} yr.
Uranium Y	90	Th^{231}	β	25.6 hr.
Protoactinium	91	Pa^{231}	α	3.248×10^{4} yr.
Actinium	89	Ac^{227}	β	21.2 yr.
Radioactinium	90	Th^{227}	α	18.17 days
Actinium X	88	Ra^{223}	α	11.7 days
Actinon (Emanation)	86	Rn^{219}	α	4.0 sec.
Actinium A	84	Po^{215}	α	1.8×10^{-3} sec.
Actinium B	82	Pb^{211}	β	36.1 min.
Actinium C	83	Bi^{211}	$\alpha + \beta$	2.15 min.
↓β Actinium C′ (↓α Actinium C″)	84	Po^{211}	α	0.52 sec.
Actinium C″	81	Tl^{207}	β	4.78 min.
↓β (from C″), ↓α (from C′) Actinium D	82	Pb^{207}	stable	—

List of Journal Abbreviations

American Journal of Science	*Amer. J. Sci.*
Annalen der Physik (Leipzig)	*Ann. Phys., Lpz*
Anzeiger der kaiserlichen Akademie der Wissenschaften. Mathematisch-naturwissenschaftliche Klasse	*Anz. Akad. Wiss., Wien*
Berichte der deutschen chemischen Gesellschaft	*Ber. dtsch chem. Ges.*
The Chemical News	*Chem. News*
Comptes Rendus de l'Académie des Sciences (Paris)	*C. R. Acad. Sci., Paris*
Jahrbuch der Radioaktivität und Elektronik	*Jb. Radioakt.*
Journal of the American Chemical Society	*J. Amer. chem. Soc.*
Journal of the Chemical Society	*J. chem. Soc.*
Nature (London)	*Nature, Lond.*
Die Naturwissenschaften	*Naturw.*
The London, Edinburgh, and Dublin Philosophical Magazine and Journal of Science	*Phil. Mag.*
Physikalische Zeitschrift	*Phys. Z.*
Proceedings of the Physical Society of London	*Proc. phys. Soc., Lond.*
Proceedings of the Royal Society (London)	*Proc. Roy. Soc., Lond.*
Sitzungsberichte der kaiserlichen Akademie der Wissenschaften in Wien. Mathematisch-naturwissenschaftliche Klasse	*S. B. Akad. Wiss., Wien*
Verhandlungen der deutschen physikalischen Gesellschaft	*Verh. dtsch phys. Ges.*
Zeitschrift für Elektrochemie	*Z. Elektrochem.*
Journal of theAmerican Chemical Society	*J. Amer. chem. Soc.*

Indexes

Numbers in **boldface** give the pages on which the reprinted papers are located. Numbers in ordinary type refer to names or topics contained within the reprinted papers. Numbers inside square brackets refer to names or topics contained in the historical essay (pp. 3–60) or the editorial commentary. *Italic* numbers indicate that a name occurs within bibliographical references cited by the editor. Those who may consider using the Index of Names as a research tool should bear in mind that it is organized as a citation index rather than a bibliography. A minor paper near the opening of a research period may be listed there more frequently than a major paper near the close.

A. Index of Names

B. Index of Topics

SOME DOVER SCIENCE BOOKS

SOME DOVER SCIENCE BOOKS

WHAT IS SCIENCE?,
Norman Campbell
This excellent introduction explains scientific method, role of mathematics, types of scientific laws. Contents: 2 aspects of science, science & nature, laws of science, discovery of laws, explanation of laws, measurement & numerical laws, applications of science. 192pp. 5⅜ x 8. 60043-2 Paperbound $1.25

FADS AND FALLACIES IN THE NAME OF SCIENCE,
Martin Gardner
Examines various cults, quack systems, frauds, delusions which at various times have masqueraded as science. Accounts of hollow-earth fanatics like Symmes; Velikovsky and wandering planets; Hoerbiger; Bellamy and the theory of multiple moons; Charles Fort; dowsing, pseudoscientific methods for finding water, ores, oil. Sections on naturopathy, iridiagnosis, zone therapy, food fads, etc. Analytical accounts of Wilhelm Reich and orgone sex energy; L. Ron Hubbard and Dianetics; A. Korzybski and General Semantics; many others. Brought up to date to include Bridey Murphy, others. Not just a collection of anecdotes, but a fair, reasoned appraisal of eccentric theory. Formerly titled *In the Name of Science*. Preface. Index. x + 384pp. 5⅜ x 8.
20394-8 Paperbound $2.00

PHYSICS, THE PIONEER SCIENCE,
L. W. Taylor
First thorough text to place all important physical phenomena in cultural-historical framework; remains best work of its kind. Exposition of physical laws, theories developed chronologically, with great historical, illustrative experiments diagrammed, described, worked out mathematically. Excellent physics text for self-study as well as class work. Vol. 1: Heat, Sound: motion, acceleration, gravitation, conservation of energy, heat engines, rotation, heat, mechanical energy, etc. 211 illus. 407pp. 5⅜ x 8. Vol. 2: Light, Electricity: images, lenses, prisms, magnetism, Ohm's law, dynamos, telegraph, quantum theory, decline of mechanical view of nature, etc. Bibliography. 13 table appendix. Index. 551 illus. 2 color plates. 508pp. 5⅜ x 8.
60565-5, 60566-3 Two volume set, paperbound $5.50

THE EVOLUTION OF SCIENTIFIC THOUGHT FROM NEWTON TO EINSTEIN,
A. d'Abro
Einstein's special and general theories of relativity, with their historical implications, are analyzed in non-technical terms. Excellent accounts of the contributions of Newton, Riemann, Weyl, Planck, Eddington, Maxwell, Lorentz and others are treated in terms of space and time, equations of electromagnetics, finiteness of the universe, methodology of science. 21 diagrams. 482pp. 5⅜ x 8.
20002-7 Paperbound $2.50

CHANCE, LUCK AND STATISTICS: THE SCIENCE OF CHANCE,
Horace C. Levinson
Theory of probability and science of statistics in simple, non-technical language. Part I deals with theory of probability, covering odd superstitions in regard to "luck," the meaning of betting odds, the law of mathematical expectation, gambling, and applications in poker, roulette, lotteries, dice, bridge, and other games of chance. Part II discusses the misuse of statistics, the concept of statistical probabilities, normal and skew frequency distributions, and statistics applied to various fields—birth rates, stock speculation, insurance rates, advertising, etc. "Presented in an easy humorous style which I consider the best kind of expository writing," Prof. A. C. Cohen, Industry Quality Control. Enlarged revised edition. Formerly titled *The Science of Chance*. Preface and two new appendices by the author. xiv + 365pp. 5⅜ x 8. 21007-3 Paperbound $2.00

BASIC ELECTRONICS,
prepared by the U.S. Navy Training Publications Center
A thorough and comprehensive manual on the fundamentals of electronics. Written clearly, it is equally useful for self-study or course work for those with a knowledge of the principles of basic electricity. Partial contents: Operating Principles of the Electron Tube; Introduction to Transistors; Power Supplies for Electronic Equipment; Tuned Circuits; Electron-Tube Amplifiers; Audio Power Amplifiers; Oscillators; Transmitters; Transmission Lines; Antennas and Propagation; Introduction to Computers; and related topics. Appendix. Index. Hundreds of illustrations and diagrams. vi + 471pp. 6½ x 9¼.
61076-4 Paperbound $2.95

BASIC THEORY AND APPLICATION OF TRANSISTORS,
prepared by the U.S. Department of the Army
An introductory manual prepared for an army training program. One of the finest available surveys of theory and application of transistor design and operation. Minimal knowledge of physics and theory of electron tubes required. Suitable for textbook use, course supplement, or home study. Chapters: Introduction; fundamental theory of transistors; transistor amplifier fundamentals; parameters, equivalent circuits, and characteristic curves; bias stabilization; transistor analysis and comparison using characteristic curves and charts; audio amplifiers; tuned amplifiers; wide-band amplifiers; oscillators; pulse and switching circuits; modulation, mixing, and demodulation; and additional semiconductor devices. Unabridged, corrected edition. 240 schematic drawings, photographs, wiring diagrams, etc. 2 Appendices. Glossary. Index. 263pp. 6½ x 9¼. 60380-6 Paperbound $1.75

GUIDE TO THE LITERATURE OF MATHEMATICS AND PHYSICS,
N. G. Parke III
Over 5000 entries included under approximately 120 major subject headings of selected most important books, monographs, periodicals, articles in English, plus important works in German, French, Italian, Spanish, Russian (many recently available works). Covers every branch of physics, math, related engineering. Includes author, title, edition, publisher, place, date, number of volumes, number of pages. A 40-page introduction on the basic problems of research and study provides useful information on the organization and use of libraries, the psychology of learning, etc. This reference work will save you hours of time. 2nd revised edition. Indices of authors, subjects, 464pp. 5⅜ x 8.
60447-0 Paperbound $2.75

COLLEGE ALGEBRA, *H. B. Fine*
Standard college text that gives a systematic and deductive structure to algebra; comprehensive, connected, with emphasis on theory. Discusses the commutative, associative, and distributive laws of number in unusual detail, and goes on with undetermined coefficients, quadratic equations, progressions, logarithms, permutations, probability, power series, and much more. Still most valuable elementary-intermediate text on the science and structure of algebra. Index. 1560 problems, all with answers. x + 631pp. 5⅜ x 8. 60211-7 Paperbound $2.75

HIGHER MATHEMATICS FOR STUDENTS OF CHEMISTRY AND PHYSICS, *J. W. Mellor*
Not abstract, but practical, building its problems out of familiar laboratory material, this covers differential calculus, coordinate, analytical geometry, functions, integral calculus, infinite series, numerical equations, differential equations, Fourier's theorem, probability, theory of errors, calculus of variations, determinants. "If the reader is not familiar with this book, it will repay him to examine it," *Chem. & Engineering News.* 800 problems. 189 figures. Bibliography. xxi + 641pp. 5⅜ x 8. 60193-5 Paperbound $3.50

TRIGONOMETRY REFRESHER FOR TECHNICAL MEN, *A. A. Klaf*
A modern question and answer text on plane and spherical trigonometry. Part I covers plane trigonometry: angles, quadrants, trigonometrical functions, graphical representation, interpolation, equations, logarithms, solution of triangles, slide rules, etc. Part II discusses applications to navigation, surveying, elasticity, architecture, and engineering. Small angles, periodic functions, vectors, polar coordinates, De Moivre's theorem, fully covered. Part III is devoted to spherical trigonometry and the solution of spherical triangles, with applications to terrestrial and astronomical problems. Special time-savers for numerical calculation. 913 questions answered for you! 1738 problems; answers to odd numbers. 494 figures. 14 pages of functions, formulae. Index. x + 629pp. 5⅜ x 8.
20371-9 Paperbound $3.00

CALCULUS REFRESHER FOR TECHNICAL MEN, *A. A. Klaf*
Not an ordinary textbook but a unique refresher for engineers, technicians, and students. An examination of the most important aspects of differential and integral calculus by means of 756 key questions. Part I covers simple differential calculus: constants, variables, functions, increments, derivatives, logarithms, curvature, etc. Part II treats fundamental concepts of integration: inspection, substitution, transformation, reduction, areas and volumes, mean value, successive and partial integration, double and triple integration. Stresses practical aspects! A 50 page section gives applications to civil and nautical engineering, electricity, stress and strain, elasticity, industrial engineering, and similar fields. 756 questions answered. 556 problems; solutions to odd numbers. 36 pages of constants, formulae. Index. v + 431pp. 5⅜ x 8. 20370-0 Paperbound $2.25

INTRODUCTION TO THE THEORY OF GROUPS OF FINITE ORDER, *R. Carmichael*
Examines fundamental theorems and their application. Beginning with sets, systems, permutations, etc., it progresses in easy stages through important types of groups: Abelian, prime power, permutation, etc. Except 1 chapter where matrices are desirable, no higher math needed. 783 exercises, problems. Index. xvi + 447pp. 5⅜ x 8. 60300-8 Paperbound $3.00

AN INTRODUCTION TO THE GEOMETRY OF N DIMENSIONS,
D. H. Y. Sommerville
An introduction presupposing no prior knowledge of the field, the only book in English devoted exclusively to higher dimensional geometry. Discusses fundamental ideas of incidence, parallelism, perpendicularity, angles between linear space; enumerative geometry; analytical geometry from projective and metric points of view; polytopes; elementary ideas in analysis situs; content of hyper-spacial figures. Bibliography. Index. 60 diagrams. 196pp. 5⅜ x 8.
60494-2 Paperbound $1.50

ELEMENTARY CONCEPTS OF TOPOLOGY, *P. Alexandroff*
First English translation of the famous brief introduction to topology for the beginner or for the mathematician not undertaking extensive study. This unusually useful intuitive approach deals primarily with the concepts of complex, cycle, and homology, and is wholly consistent with current investigations. Ranges from basic concepts of set-theoretic topology to the concept of Betti groups. "Glowing example of harmony between intuition and thought," David Hilbert. Translated by A. E. Farley. Introduction by D. Hilbert. Index. 25 figures. 73pp. 5⅜ x 8. 60747-X Paperbound $1.25

ELEMENTS OF NON-EUCLIDEAN GEOMETRY,
D. M. Y. Sommerville
Unique in proceeding step-by-step, in the manner of traditional geometry. Enables the student with only a good knowledge of high school algebra and geometry to grasp elementary hyperbolic, elliptic, analytic non-Euclidean geometries; space curvature and its philosophical implications; theory of radical axes; homothetic centres and systems of circles; parataxy and parallelism; absolute measure; Gauss' proof of the defect area theorem; geodesic representation; much more, all with exceptional clarity. 126 problems at chapter endings provide progressive practice and familiarity. 133 figures. Index. xvi + 274pp. 5⅜ x 8. 60460-8 Paperbound $2.00

INTRODUCTION TO THE THEORY OF NUMBERS, *L. E. Dickson*
Thorough, comprehensive approach with adequate coverage of classical literature, an introductory volume beginners can follow. Chapters on divisibility, congruences, quadratic residues & reciprocity. Diophantine equations, etc. Full treatment of binary quadratic forms without usual restriction to integral coefficients. Covers infinitude of primes, least residues. Fermat's theorem. Euler's phi function, Legendre's symbol, Gauss's lemma, automorphs, reduced forms, recent theorems of Thue & Siegel, many more. Much material not readily available elsewhere. 239 problems. Index. I figure. viii + 183pp. 5⅜ x 8.
60342-3 Paperbound $1.75

MATHEMATICAL TABLES AND FORMULAS,
compiled by Robert D. Carmichael and Edwin R. Smith
Valuable collection for students, etc. Contains all tables necessary in college algebra and trigonometry, such as five-place common logarithms, logarithmic sines and tangents of small angles, logarithmic trigonometric functions, natural trigonometric functions, four-place antilogarithms, tables for changing from sexagesimal to circular and from circular to sexagesimal measure of angles, etc. Also many tables and formulas not ordinarily accessible, including powers, roots, and reciprocals, exponential and hyperbolic functions, ten-place logarithms of prime numbers, and formulas and theorems from analytical and elementary geometry and from calculus. Explanatory introduction. viii + 269pp. 5⅜ x 8½. 60111-0 Paperbound $1.50

THE RISE OF THE NEW PHYSICS (formerly THE DECLINE OF MECHANISM), *A. d'Abro*
This authoritative and comprehensive 2-volume exposition is unique in scientific publishing. Written for intelligent readers not familiar with higher mathematics, it is the only thorough explanation in non-technical language of modern mathematical-physical theory. Combining both history and exposition, it ranges from classical Newtonian concepts up through the electronic theories of Dirac and Heisenberg, the statistical mechanics of Fermi, and Einstein's relativity theories. "A must for anyone doing serious study in the physical sciences," *J. of Franklin Inst.* 97 illustrations. 991pp. 2 volumes.
20003-5, 20004-3 Two volume set, paperbound \$5.50

THE STRANGE STORY OF THE QUANTUM, AN ACCOUNT FOR THE GENERAL READER OF THE GROWTH OF IDEAS UNDERLYING OUR PRESENT ATOMIC KNOWLEDGE, *B. Hoffmann*
Presents lucidly and expertly, with barest amount of mathematics, the problems and theories which led to modern quantum physics. Dr. Hoffmann begins with the closing years of the 19th century, when certain trifling discrepancies were noticed, and with illuminating analogies and examples takes you through the brilliant concepts of Planck, Einstein, Pauli, de Broglie, Bohr, Schroedinger, Heisenberg, Dirac, Sommerfeld, Feynman, etc. This edition includes a new, long postscript carrying the story through 1958. "Of the books attempting an account of the history and contents of our modern atomic physics which have come to my attention, this is the best," H. Margenau, Yale University, in *American Journal of Physics.* 32 tables and line illustrations. Index. 275pp. 5⅜ x 8.
20518-5 Paperbound \$2.00

GREAT IDEAS AND THEORIES OF MODERN COSMOLOGY, *Jagjit Singh*
The theories of Jeans, Eddington, Milne, Kant, Bondi, Gold, Newton, Einstein, Gamow, Hoyle, Dirac, Kuiper, Hubble, Weizsäcker and many others on such cosmological questions as the origin of the universe, space and time, planet formation, "continuous creation," the birth, life, and death of the stars, the origin of the galaxies, etc. By the author of the popular *Great Ideas of Modern Mathematics.* A gifted popularizer of science, he makes the most difficult abstractions crystal-clear even to the most non-mathematical reader. Index. xii + 276pp. 5⅜ x 8½. 20925-3 Paperbound \$2.50

GREAT IDEAS OF MODERN MATHEMATICS: THEIR NATURE AND USE, *Jagjit Singh*
Reader with only high school math will understand main mathematical ideas of modern physics, astronomy, genetics, psychology, evolution, etc., better than many who use them as tools, but comprehend little of their basic structure. Author uses his wide knowledge of non-mathematical fields in brilliant exposition of differential equations, matrices, group theory, logic, statistics, problems of mathematical foundations, imaginary numbers, vectors. etc. Original publications, appendices. indexes. 65 illustr. 322pp. 5⅜ x 8. 20587-8 Paperbound \$2.25

THE MATHEMATICS OF GREAT AMATEURS, *Julian L. Coolidge*
Great discoveries made by poets, theologians, philosophers, artists and other non-mathematicians: Omar Khayyam, Leonardo da Vinci, Albrecht Dürer, John Napier, Pascal, Diderot, Bolzano, etc. Surprising accounts of what can result from a non-professional preoccupation with the oldest of sciences. 56 figures. viii + 211pp. 5⅜ x 8½. 61009-8 Paperbound \$2.00

FIVE VOLUME "THEORY OF FUNCTIONS" SET BY KONRAD KNOPP

This five-volume set, prepared by Konrad Knopp, provides a complete and readily followed account of theory of functions. Proofs are given concisely, yet without sacrifice of completeness or rigor. These volumes are used as texts by such universities as M.I.T., University of Chicago, N. Y. City College, and many others. "Excellent introduction . . . remarkably readable, concise, clear, rigorous," *Journal of the American Statistical Association.*

ELEMENTS OF THE THEORY OF FUNCTIONS,
Konrad Knopp
This book provides the student with background for further volumes in this set, or texts on a similar level. Partial contents: foundations, system of complex numbers and the Gaussian plane of numbers, Riemann sphere of numbers, mapping by linear functions, normal forms, the logarithm, the cyclometric functions and binomial series. "Not only for the young student, but also for the student who knows all about what is in it," *Mathematical Journal.* Bibliography. Index. 140pp. 5⅜ x 8. 60154-4 Paperbound $1.50

THEORY OF FUNCTIONS, PART I,
Konrad Knopp
With volume II, this book provides coverage of basic concepts and theorems. Partial contents: numbers and points, functions of a complex variable, integral of a continuous function, Cauchy's integral theorem, Cauchy's integral formulae, series with variable terms, expansion of analytic functions in power series, analytic continuation and complete definition of analytic functions, entire transcendental functions, Laurent expansion, types of singularities. Bibliography. Index. vii + 146pp. 5⅜ x 8. 60156-0 Paperbound $1.50

THEORY OF FUNCTIONS, PART II,
Konrad Knopp
Application and further development of general theory, special topics. Single valued functions. Entire, Weierstrass, Meromorphic functions. Riemann surfaces. Algebraic functions. Analytical configuration, Riemann surface. Bibliography. Index. x + 150pp. 5⅜ x 8. 60157-9 Paperbound $1.50

PROBLEM BOOK IN THE THEORY OF FUNCTIONS, VOLUME 1.
Konrad Knopp
Problems in elementary theory, for use with Knopp's *Theory of Functions,* or any other text, arranged according to increasing difficulty. Fundamental concepts, sequences of numbers and infinite series, complex variable, integral theorems, development in series, conformal mapping. 182 problems. Answers. viii + 126pp. 5⅜ x 8. 60158-7 Paperbound $1.50

PROBLEM BOOK IN THE THEORY OF FUNCTIONS, VOLUME 2,
Konrad Knopp
Advanced theory of functions, to be used either with Knopp's *Theory of Functions,* or any other comparable text. Singularities, entire & meromorphic functions, periodic, analytic, continuation, multiple-valued functions, Riemann surfaces, conformal mapping. Includes a section of additional elementary problems. "The difficult task of selecting from the immense material of the modern theory of functions the problems just within the reach of the beginner is here masterfully accomplished," *Am. Math. Soc.* Answers. 138pp. 5⅜ x 8.
60159-5 Paperbound $1.50

NUMERICAL SOLUTIONS OF DIFFERENTIAL EQUATIONS,
H. Levy & E. A. Baggott
Comprehensive collection of methods for solving ordinary differential equations of first and higher order. All must pass 2 requirements: easy to grasp and practical, more rapid than school methods. Partial contents: graphical integration of differential equations, graphical methods for detailed solution. Numerical solution. Simultaneous equations and equations of 2nd and higher orders. "Should be in the hands of all in research in applied mathematics, teaching," *Nature*. 21 figures. viii + 238pp. 5⅜ x 8. 60168-4 Paperbound $1.85

ELEMENTARY STATISTICS, WITH APPLICATIONS IN MEDICINE AND THE BIOLOGICAL SCIENCES, *F. E. Croxton*
A sound introduction to statistics for anyone in the physical sciences, assuming no prior acquaintance and requiring only a modest knowledge of math. All basic formulas carefully explained and illustrated; all necessary reference tables included. From basic terms and concepts, the study proceeds to frequency distribution, linear, non-linear, and multiple correlation, skewness, kurtosis, etc. A large section deals with reliability and significance of statistical methods. Containing concrete examples from medicine and biology, this book will prove unusually helpful to workers in those fields who increasingly must evaluate, check, and interpret statistics. Formerly titled "Elementary Statistics with Applications in Medicine." 101 charts. 57 tables. 14 appendices. Index. vi + 376pp. 5⅜ x 8. 60506-X Paperbound $2.25

INTRODUCTION TO SYMBOLIC LOGIC,
S. Langer
No special knowledge of math required — probably the clearest book ever written on symbolic logic, suitable for the layman, general scientist, and philosopher. You start with simple symbols and advance to a knowledge of the Boole-Schroeder and Russell-Whitehead systems. Forms, logical structure, classes, the calculus of propositions, logic of the syllogism, etc. are all covered. "One of the clearest and simplest introductions," *Mathematics Gazette*. Second enlarged, revised edition. 368pp. 5⅜ x 8. 60164-1 Paperbound $2.25

A SHORT ACCOUNT OF THE HISTORY OF MATHEMATICS,
W. W. R. Ball
Most readable non-technical history of mathematics treats lives, discoveries of every important figure from Egyptian, Phoenician, mathematicians to late 19th century. Discusses schools of Ionia, Pythagoras, Athens, Cyzicus, Alexandria, Byzantium, systems of numeration; primitive arithmetic; Middle Ages, Renaissance, including Arabs, Bacon, Regiomontanus, Tartaglia, Cardan, Stevinus, Galileo, Kepler; modern mathematics of Descartes, Pascal, Wallis, Huygens, Newton, Leibnitz, d'Alembert, Euler, Lambert, Laplace, Legendre, Gauss, Hermite, Weierstrass, scores more. Index. 25 figures. 546pp. 5⅜ x 8.
20630-0 Paperbound $2.75

INTRODUCTION TO NONLINEAR DIFFERENTIAL AND INTEGRAL EQUATIONS,
Harold T. Davis
Aspects of the problem of nonlinear equations, transformations that lead to equations solvable by classical means, results in special cases, and useful generalizations. Thorough, but easily followed by mathematically sophisticated reader who knows little about non-linear equations. 137 problems for student to solve. xv + 566pp. 5⅜ x 8½. 60971-5 Paperbound $2.75

A SOURCE BOOK IN MATHEMATICS,
D. E. Smith
Great discoveries in math, from Renaissance to end of 19th century, in English translation. Read announcements by Dedekind, Gauss, Delamain, Pascal, Fermat, Newton, Abel, Lobachevsky, Bolyai, Riemann, De Moivre, Legendre, Laplace, others of discoveries about imaginary numbers, number congruence, slide rule, equations, symbolism, cubic algebraic equations, non-Euclidean forms of geometry, calculus, function theory, quaternions, etc. Succinct selections from 125 different treatises, articles, most unavailable elsewhere in English. Each article preceded by biographical introduction. Vol. I: Fields of Number, Algebra. Index. 32 illus. 338pp. 5⅜ x 8. Vol. II: Fields of Geometry, Probability, Calculus, Functions, Quaternions. 83 illus. 432pp. 5⅜ x 8.
60552-3, 60553-1 Two volume set, paperbound $5.00

FOUNDATIONS OF PHYSICS,
R. B. Lindsay & H. Margenau
Excellent bridge between semi-popular works & technical treatises. A discussion of methods of physical description, construction of theory; valuable for physicist with elementary calculus who is interested in ideas that give meaning to data, tools of modern physics. Contents include symbolism; mathematical equations; space & time foundations of mechanics; probability; physics & continua; electron theory; special & general relativity; quantum mechanics; causality. "Thorough and yet not overdetailed. Unreservedly recommended," *Nature* (London). Unabridged, corrected edition. List of recommended readings. 35 illustrations. xi + 537pp. 5⅜ x 8. 60377-6 Paperbound $3.50

FUNDAMENTAL FORMULAS OF PHYSICS,
ed. by D. H. Menzel
High useful, full, inexpensive reference and study text, ranging from simple to highly sophisticated operations. Mathematics integrated into text—each chapter stands as short textbook of field represented. Vol. 1: Statistics, Physical Constants, Special Theory of Relativity, Hydrodynamics, Aerodynamics, Boundary Value Problems in Math, Physics, Viscosity, Electromagnetic Theory, etc. Vol. 2: Sound, Acoustics, Geometrical Optics, Electron Optics, High-Energy Phenomena, Magnetism, Biophysics, much more. Index. Total of 800pp. 5⅜ x 8.
60595-7, 60596-5 Two volume set, paperbound $4.75

THEORETICAL PHYSICS,
A. S. Kompaneyets
One of the very few thorough studies of the subject in this price range. Provides advanced students with a comprehensive theoretical background. Especially strong on recent experimentation and developments in quantum theory. Contents: Mechanics (Generalized Coordinates, Lagrange's Equation, Collision of Particles, etc.), Electrodynamics (Vector Analysis, Maxwell's equations, Transmission of Signals, Theory of Relativity, etc.), Quantum Mechanics (the Inadequacy of Classical Mechanics, the Wave Equation, Motion in a Central Field, Quantum Theory of Radiation, Quantum Theories of Dispersion and Scattering, etc.), and Statistical Physics (Equilibrium Distribution of Molecules in an Ideal Gas, Boltzmann Statistics, Bose and Fermi Distribution. Thermodynamic Quantities, etc.). Revised to 1961. Translated by George Yankovsky, authorized by Kompaneyets. 137 exercises. 56 figures. 529pp. 5⅜ x 8½.
60972-3 Paperbound $3.50